ROMEO & THE ORPHAN OF HAMLET

Prologue:

No one is ever who others think they are. We are each and every one of us a unique individual. We are able to hide ourselves from each other within a culture of education and experiences.

There are three things in our lives that give us a depth of mind, where we know who we really are. This story is about being and becoming.

I0469697

CHAPTER ONE

OZ PARK CHICAGO

Romeo and Roberto hurried off to their chemistry class. Professor Periodicus scowled as they entered the room just as the bell was ringing. Both boys flopped down into their seats.

"Perhaps you two could arrive at least five minutes earlier so that we can begin on time?" He asked.

"Sorry Mr. Periodicus." Roberto told him.

"It is Professor Periodicus to you Roberto!"

"Yes sir. We were delayed by construction workers in Oz Park who were pouring cement."

"They made us stand and watch while the cement truck blocked our way here. Professor Periodicus." Romeo added.

"Honest Professor, we were not allowed to pass until the cement truck left the scene." Roberto explained. "We will not take that shortcut again."

The professor raised his finger and pointed to both boys. "Not unless you two return to write your names in the wet cement. Do you think I don't know what boys your age, do?"

Before Romeo and Roberto could reply, another student stood up and spoke.

"Excuse me Professor. Is this a class teaching us about chemistry or is it about Romeo and Roberto writing in wet cement?"

The professor thought for one moment. He had already wasted ten minutes presenting his lecture on carbon as the basis of all living chemical elements in the known universe. The student who asked the question was challenging him. The Professor dismissed the class.

"Winslow! You will meet with me in my office with Romeo and Roberto."

MERRIMAC ACADEMY OF MATHEMATICS AND SCIENCE

Headmaster Applebee was chosen to administer a complex educational unit of students who were what his superiors labeled exceptional. His qualifications were simple. He was to love his job more than understand it.

Once Professor Periodicus explained his dilemma with Romeo, Roberto and Winslow in the chemistry class, Headmaster Applebee asked the Professor to sit next to him on his plush sofa. Applebee did not believe in having a desk, as a power barrier, between him and those who entered his office.

Professor Periodicus sat down. Swoosh.

"Who are you referring to?" The Headmaster asked.

"They are students in my chemistry class sir, Romeo and Roberto. Winslow stood up for them when I attempted to discipline them for their tardiness."

"According to my conversation with Winslow, they arrived just as the bell was ringing. We understand that instead of presenting your lecture, you choose to spend the time debating the subject of wet cement?"

"My teaching methods are not negotiable. It is time for you to discipline those students!"

"You mean to say Professor Periodicus that you choose to end your lecture in order to teach discipline instead of chemistry?"

"Who runs this school," the Professor asked, "you or the students?"

"At Merrimac Academy the students select their teachers and their classes." Headmaster Applebee told him. Your contract requires you to have students if you wish to continue teaching at Merrimac."

"What does this have to do with Winslow and the other two students?"

PROFESSOR PERIODICUS'S OFFICE

The three students entered the office. "You wanted to see us Professor?" Winslow asked.

"So you are the spokesman in this group. Take a seat across from my desk. The sofa is reserved for those students who need help with their chemistry studies. I want them to be comfortable while they seek help from me. The chairs are for students who interrupt my lectures."

"Students who interrupt your lectures need help also," Romeo told him, "so why can't we sit on your sofa?"

"Are you trying to hide something from us, like how small your penis is?" Roberto asked.

"You have already shown us how big your ego is and the size of your pea shaped brain." Winslow told him.

"Once the Headmaster is informed of your comments, you will all be suspended, if not expelled. My Doctorate, PHD to you three ruffians, will crush all of you!"

"We are truly sorry Professor Periodicus. We did not know that a PHD could be used to torture and crush students who have found out that you are a phony, hypocrite and a deviate who enjoys sitting on couches with his male students." Winslow apologized.

"I have never abused one of my students!" The Professor exclaimed.

Winslow sat on the professor's sofa. "That is because sir, they were able to zip up their pants faster than you could unzip them. Every student here knows that if they want a perfect score, in chemistry at the end of the semester, they had only to unbuckle their belt on your couch."

Professor Periodicus had a PHD but he was not a dummy. He had the right to change one of his student's final grade score from an A to an F.

Winslow was joined by Romeo and Roberto on the Professor's sofa.

"What your PHD neglected to explain to you Professor is that the higher the degree, the warmer the temperature. It is only a matter of time before the pot boils over, spilling the truth of who you really are. You are not true to yourself. You are here, in this Academy, among the smartest boys in Chicago, who not only seek knowledge of your chemistry but of life itself. They do not need you to seduce them, on your sofa, in order to graduate from Merrimac. They want to discover another dimension of what is that happening in our time."

"Does this include us?" Roberto asked.

"Both you and Romeo and myself," Winslow explained.

Professor Periodicus sat in his special chair behind the desk. His knowledge of chemical elements, as well as his apt skills in finding ways to lure others, especially young boys into having sex with him was easy. He discovered that 12 to 13 year old boys were the hottest and horniest of the male generation. Their testicles could not produce enough sperm for their masturbations and wet dreams.

"Then why are we here in Professor Periodicus's office? We didn't do anything wrong." Roberto asked.

"No, but he used your being in class at the last moment as an excuse to have you both sit on his office sofa," Winslow explained.

Romeo and Roberto looked at each other. They were confused. Winslow left the room. Professor Periodicus told Romero and Roberto to leave also.

HEADMASTER APPLEBEES OFFICE

"Those three students elected to attend your chemistry lectures and labs. They were interviewed by me this morning after you filed a disciplinary report on them. Did you think Professor Perodicus that I would not hear their side of what took place in your class? Are you aware that Romeo taped the conversation when you summoned them to your office? Romeos father graduated from this academy with high honors. His uncle is a lawyer who supplied him with the recording device. Winslow's father is on the board of directors."

"What about the other smart mouthed boy?" The Professor asked.

"Romeo's parents are the guardians of Roberto who lives with them on Taylor Street." Applebee told him.

Professor Periodicus was not backing down on his disciplinary complaint. "You cannot believe that these boys are telling you the truth. They are failing my class! Why should you believe them? I have never abused one of our students. No one can prove otherwise. They must be punished for bringing these accusations to you. The rest of my students will tell you otherwise. It is illegal to tape a conversation without my permission."

"No professor. It cannot be used as evidence in a court of law. It is not illegal to tape anyone's conversation unless it is used for other illegal purposes. The law has many elements covering the subject, not unlike your periodic charts."

"The rest of my students will register for the next semester."

"Unfortunately for you Professor not one of our students has elected to study with you next semester. The Board of Directors has placed you on an extended sabbatical leave without pay." The Headmaster explained to the Professor.

"That is a fine way to run an Academy Applebee! Who wants to teach students who are able to choose their teachers?"

"You would be surprised to find that those who seek knowledge succeed with teachers who challenge them beyond all expectations."

"Expectations of what," The professor asked, "students know nothing of what they expect from a teacher."

"Let us start with what they do not expect Professor. Every student thirsting for knowledge is also seeking truth. From birth to death all of us search for something more. What drove anyone to discover the elements of the universe, or how to build a pyramid?

Who put together the ancient temples of the Greeks, the Incas or, the Mayans? How was it possible to rivet steel plates together and replace wooden ships with steamer ships?

Expecting nothing from their teacher, the student languishes and surrenders any hope of finding the reason to learn. Do you really think our students do not expect your faith and hope in the future, to rub off on them?"

"They are not the scholars you want them to become in this Academy Applebee! Your students come from privileged society."

"So Professor, you believe that a boy from the ghetto is not able to transcend the adverse environment he is born into?"

"Students are students, nothing more than a means to an end. They are the reason why teachers exist."

Headmaster Applebee left the comfort of his sofa. "How did you ever end up teaching in Merrimac with your credentials?"

"Easy Applebee, you neglected to search my background in order to discover that my named was changed. Anyone can forge a new passport or driver ID. You did not go far enough into my background to find out that I was listed on the federal NCIC as a sex offender."

"We trusted you. Your academic credentials were excellent. Why did you deceive us?"

"This was a good place to be. Someone asked Jesse James why he robbed banks. His reply was, because that is where the money is. Merrimac was where the boys were. The students I had sex with were boys who did not have a father living with them. They wanted a man to love them. They did not want sex, at first, until I touched one of them as he sat on my sofa. We would smoke a pipe and blow smoke at each other. We blew the smoke into our ears and mouth while sitting together. They always came back for more. Finally we touched each other below the belt. I would blow smoke into his hair until one day he reached over and placed his hand on my knee. I responded by lifting his tee shirt and softly touching his chest, I kissed the boy on the head and told him that I loved him. He asked me to take him home. It was the first time that I fell in love with a boy. He was only fourteen. We never saw each other after that, once I dropped him off at his house."

SCARECROW OZ PARK

Uncle Patrick did not like meeting his nephew and Roberto near the statue on Webster Street in OZ Park. He did not mind dropping the boys off in the morning on his way to the courthouse but they were never on time when he arrived to escort them home.

"Sorry we are late." Romero said as they ran up to him. Our chemistry professor had some kind of a problem with another student."

"His name is Winslow. We had to go with him to the professor's office. There was some kind of a problem about sitting on his sofa. We have a new friend at school now." Roberto explained.

"I heard. The Headmaster called your mother Romeo." His uncle told him.

"Are we in trouble?" Roberto asked. "The Professor was angry at what our friend said to him."

"We didn't do anything wrong." Romeo exclaimed.

"Let me get you both back home. You can explain it to your mother and father tonight."

"Do we need a lawyer?" Romeo asked.

His uncle laughed. "Maybe you two hooligans can tell me the story on the way."

CHAPTER TWO

THE NEW FRIEND

Winslow was seated on the far side of the school cafeteria. Romeo spotted him as he and Roberto paid for their lunch. "Look Roberto. Isn't Winslow sitting over there?"

"Let's join him." Roberto suggested.

"Maybe it's not a good idea considering what happened yesterday in Periodicus's office."

"Remember when you asked me who was going to take us to the next dimension. You told me that someone in this academy must know how to get there." Roberto reminded him.

"When we pursue the unknown and the strange, we discover what makes us what we are. That is what I said Roberto."

"Then you told me we must find the person we are looking for, or they must discover us." Roberto added.

"Follow me." Romeo told him. "We told Uncle Patrick that Winslow was our new friend."

As they sat down across from Winslow, they waited for him to speak first. "If it isn't the cement boys who were brave enough to sit on the sofa with me! It took a lot of courage for you to follow my lead."

"My mother received a call from the Headmaster." Romeo told him.

"As did my own father who solved the whole problem by suggesting that the professor be placed on leave without pay, until the problem is resolved." Winslow told them.

"But what was the problem?" Roberto asked "We didn't do anything wrong."

"The professor was the problem. He will not be lecturing in the halls of Merrimac Academy again."

Romeo's eyes lit up. "Did it have anything to do with the sofa in his office?"

"Not the sofa but what took place on the sofa with the professor and his students."

"In Italy he would be called a finochio."Roberto said.

Winslow was easy to talk to. He could easily see the bond between Romeo and Roberto. On the other hand, they could see the depth of mind in their new friend.

"You are surrounded by those who truly love and understand you. We are all on a journey' from before our time to beyond our time. Great scholars have guided you here. Both of you are seeking another dimension." Winslow said. It shows in your eyes."

"We have been looking for you but you have found us." Romeo declared, holding Roberto's hand under the table.

"Reach for stars you've never seen. Believe me they are there." Winslow promised. "We are surrounded by others in another dimension. You are cared for and protected by a father who graduated from this academy. His cousin is a graduate at the University of Illinois in Chicago. He is studying for his PHD in Classical Greek Philosophy and Literature. You two have travelled Greece and Italy this summer. Your tans are warm and beautiful. His mentor is Professor Xthanlos who shares the living quarters with your cousin Andretti on the third floor of your Taylor Street home.

Your mother's brother occupies the second floor law offices along with the rooms which you and Roberto share with each other. Your family conducts weekly fireside chats on the first floor. Your father and mother conduct a very successful business on the Southside of our windy city. You both have been looking for me and I have been looking for you."

"What if we did not choose to sit at this table with you?" Roberto asked.

"The scarecrow would never use his brain again. Notice that both of you started your day almost missing chemistry class? Then you found your lawyer uncle who sat and stared across Webster Street, while he was waiting for you to show up."

"How do you know, all of this, Winslow?" Romeo asked.

"You say that you are seeking another dimension. Do both of you have the courage to allow me to show it to you?"

Romeo grasped Roberto's hand. "Yes we can do it. Will we be able to still deal with our family?"

"The fourth dimension is a way to reach everyone. You will be able to connect with everyone who is in the third dimension."

"Yes Winslow, because when I entered the pyramid in Egypt there was a quiet moment when they spoke to me about you. Whoever they were, communicated from another dimension, saying that we would meet once we returned here to find you."

"We drifted off into another world, like waking up from a daydream." Roberto added.

"What is it like to enter the fourth dimension?" Romeo asked.

"You are in the same place at the same time with others in the third dimension except that they cannot see you. You can travel from one point in the fourth dimension to another point at twice the speed of light. It is what the physicists call a quantum leap. In quantum physics it is possible to stop the speed of light, even bending it around third dimension objects or matter."

"Where did you learn about all of this, Winslow?" Romeo asked.

"My father designed the first quantum computer."

"We are ready to meet this computer Winslow. When can you show us the next dimension?"

"We are not afraid to enter the fourth dimension." Roberto spoke for both of them.

"The first step is to meet me beneath the scarecrow. Meet me here after school tomorrow."

"My uncle Patrick is going to pick us up by the scarecrow after school. We will be grounded if we are late again like yesterday. My parents cannot know about this!"

Winslow thought a moment. Your uncle does legal work for my father who, as you know, is on the Board of Directors at Merrimac. He is able to have the Headmaster call your uncle and make an appointment with him on Taylor Street. He will offer to drive both of you home after a faculty meeting. In the meantime you both will do your homework in the library."

"Or meet you under the scarecrow and find the Quantum Computer."

"What if Uncle Patrick decides to send someone else to pick us up?"

"He will not reach anyone on his phones. Quantum Computer can control your uncle's phone systems."

"Can you do that? Isn't it a violation of privacy or the right to be free?" Roberto asked.

Romeo nodded his head in agreement. "Professor Xthanlos and Andretti told me that if we cannot control others they will control us."

"Only we provide our own privacy or our right to be." Winslow told them. "Our own constitution requires its citizens to hire and pay our way through the judicial system, for a final decision, in a Supreme Court of Justices hiding behind their black robes."

"What kind of a world awards a medal to a soldier for killing a man and gives him a dishonorable discharge for loving a man?" Roberto asked.

"It is the same world that creates war, while it enacts laws against same sex marriage." Winslow replied. War fosters major changes in human behavior. All religions have strict dogmas fighting against change. "Professor Xthanlos and your father's cousin Andretti are correct when they declare that a good philosophy is far greater than any religion."

"This is where it all began. My father and Andretti met Professor Xthanlos under the Scarecrow statue at midnight. It is right for our adventure to begin there also." Romeo told them.

"See you after school Winslow." Roberto told him.

"When you look across the park toward Webster Street, my house is on the other side. Ring the bell."

WINSLOW'S SECRET ROOM

Winslow let them into the house. The front hall was very large with a high ceiling. Statues stood along both walls leading toward a wide stairway.

Winslow took them to the bottom of the stairs. "Wait here please. Be right back." He then disappeared into one of the rooms.

"What about all of those statues we just passed?" Romeo asked. "They were all naked!"

"They were nude Romeo. The Greeks and the Romans thought the nude body was a work of art."

"But isn't it wrong to be naked unless we are in the bathroom or in bed?"

Romeo, we all are always naked underneath our clothes. "What did the scarecrow want?"

"He wanted a brain."

"Yes. And what did the tin man want Romeo?"

"He wanted a heart. But what does that have to do with…."

Roberto interrupted. "What did the lion want?"

Romeo plunked down on the steps. "He wanted courage Roberto. What do they have to do with us and Winslow?"

Roberto joined his friend on the stairs. "Don't you see it? Can't you put all of this together?"

Before he could answer, Roberto turned at the sound of Winslow's voice, speaking to them from the top of the stairs.

"Roberto is right Romeo. We are no longer in Kansas."

"What's that supposed to mean Winslow?" Romeo asked.

The two boys put one foot in front of the other until they were next to Winslow, looking down the length of the hall from above.

Romeo looked at Roberto for more courage.

"Take heart." Winslow told them. "Your brains are the reason why you are here."

Romeo and Roberto both felt that they would never be the same again, brains or no brains.

CHAPTER THREE

ROMEO AND ROBERTO MEET QUANTUM

Winslow took them down another corridor. They walked by many doors but this time there were no statues. Beautiful photographs were hung between each doorway, pictures of planets, constellations, and moons. Romeo stopped in front of one picture showing earth's moon.

"Winslow. This not the side we see from earth. It is the dark side of the moon."

"We can thank President Kennedy and NASA for this photo." Winslow told them. "This is the door to my secret room.

They entered the room closing the door behind them.

"This is a bathroom!" Romeo said.

"What is so secret about this room?" Roberto asked.

"Remember I told you it was the doorway to my secret room."

"Now what do we do?" Romeo asked.

"Put these togas on."

"You mean over our clothes?"

Winslow handed them each a toga. "No. Take off your clothes and put on your toga."

Roberto stripped and pulled his toga over his head. "Come on Romeo. We did this in Greece during the celebration dance at a wedding where they broke all of those dishes. We've come this far, so we might as well go all the way."

"Why do we need to take off all of our clothes?" Romeo asked.

"Metal objects distort the magnetic fields." Winslow explained. "The togas are itch free merino wool from Australia."

"What if we have dental fillings or gold teeth?" Roberto asked.
"You don't." Winslow told them.

Romeo looked at Winslow with suspicion. "How do you know that?"

"I saw your X-Rays in the nurse's office."

"Our medical records are private information Winslow! How did you get to see them?" Romeo challenged him.

"You are about to find out when you meet Quantum."

"The three of them stripped and put on the Togas.

"You look good in those." Winslow told them.

"They are really comfortable." Romeo agreed.

"Of course they are!" Winslow exclaimed. Your genitals are not pushed up against your bum crack. Bullies cannot give you a wedge. When you have to piss you just lift up the front of your toga. When you sit you lift up the back."

"Now what do we do?" Romeo asked.

"We climb the stairs to meet Quantum." Winslow said.

Romeo looked around the bathroom.

"What stairs?"

THE BROWNSTONE – TAYLOR STREET

The family was gathered in the drawing room waiting for Romeo and Roberto to arrive. Patrick and his wife to be, Carmen, entered. Professor Xthanlos and Andretti stood to greet her.
"So you are turning in your typing for another position?" The Professor asked. How will Patrick run his law office without you?"

"When is the wedding?" Andretti asked them. Hopefully it is before we return to Greece for the summer." They kissed the couple on both cheeks. "Perhaps you can travel with us to spend your honeymoon on the Mediterranean.

"Rita, do your parents know their son is taking the big step?" Angelo asked?

"Mother will have a lot of questions. My father will be pleased to know that there are more children on the way, maybe. I cannot speak for Patrick and Carmen."

"We already have one on the way." Carmen told them, smiling and taking Patrick's hand.

"What are you going to tell Isabelle?" Rita asked. She wanted Angelo and me to prevent Romeo's birth, telling us we would burn in hell for our sins. It didn't matter that abortion was not sanctioned by her church."

"Isabelle and Giuseppe will be here for dinner!" Rita cried. "Where are the boys Patrick?"

"They will arrive any minute. Headmaster Applebee is driving them from the school library. I have a meeting with him in my office about fifteen minutes from now."

"Applebee was one of my best teachers at the academy!" Andretti exclaimed.

Patrick excused himself as he and Carmen left the room. "See you at dinner Rita. Make sure Isabelle is well into her cups before then. Carmen and I will make our plans known before dessert. Our father can look forward to crafting a new crib in his basement workshop."

Roberto and Romeo burst into the sitting room, throwing their back packs on the floor while they ran to Angelo and Rita.

"Your snacks are in the kitchen." Rita told them, following each kiss on the cheek. "Then go to your rooms and change for dinner."

"Finish your homework before dinner." Angelo told them. Giuseppe and Isabelle will be here."

"We did our homework in the library." Romeo told his father.

Rita clapped her hands. "Good, then you can entertain your grandparents while we finish preparing dinner!"

"I will be the cocktail waiter." Roberto volunteered.

"Grandfather and I will tell jokes and riddles." Romeo said.

"But only when I am back in the room with your grandmother's libations." Roberto pleaded. "Giuseppe asks good riddles."

"How about if he tells me a riddle while you are in the kitchen and a riddle when you get back? That way you can solve the riddles you didn't hear when we go to bed."

Roberto agreed as he ran to answer the door. Giuseppe and Isabelle entered the drawing room. "Please be seated." He asked them, taking their coats. "I am your waiter tonight."

"Mother and father will be with you as soon as the dinner is in the oven. Romeo told her.

 "Meanwhile our wine steward will serve you."

"What about Uncle Patrick and that secretary of his?" Isabelle asked.

"You mean his Para-Legal assistant Carmen? They will join us when they are finished with a client." Roberto replied.

"If you boys ask me she is more than an assistant!"

"Now Isabelle, you promised not to make any scenes." Giuseppe cautioned.

"That's OK Grandfather. She didn't ask us." Romeo said.

Roberto stood straight as he clasped his hands. "What is your pleasure this evening madam? I know that sir prefers his glass of red Chianti."

Romeo stood behind Roberto and whispered. "Chianti is always red!"

"Yes but she doesn't know that." Roberto whispered back.

"I will have my usual vodka martini. This time put an olive in it, not one of those baby onions."

"As you wish madam," Roberto said retreating to the kitchen.

Giuseppe winked at Romeo. What is always coming but when it arrives it is never here?" He asked.

"Where are those two scholars who live upstairs?" Isabelle interrupted. She did not like riddles.

"They will be down for dinner grandmother. Andretti is helping Professor Xthanlos grade his student's papers on Ancient Greek Mythology. They help each other all the time." Romeo explained.

"I bet they work very closely." She scowled, as Roberto returned with the drinks. He graciously served grandmother as he held the tray. "See this," She said turning to her husband. Our waiter served his lady first."

Giuseppe took the generous goblet of wine and winked at both boys. Romeo and Roberto retreated to the other side of the room.

"Did you guess the answer to his riddle?" Roberto asked.

"Not yet. She asked me about Andretti and the Professor. She doesn't like riddles."

"What does she like Romeo?"
"The only thing I can think of is what she is sipping on. Later she will be taking larger sips. Then I will have time to think about an answer to my grandfather's riddles."

HEADMASTER APPLEBEE AND UNCLE PATRICK

Carmen greeted the Headmaster of Merrimac Academy when he entered the law office. She directed him to a sofa near a large window. "Patrick will be right with you."

The lawyer entered his office from the other side of the room. "Thank you for bringing the boys home from school and keeping your appointment at the same time."

They shook hands and sat across from each other as Patrick settled in his rocking chair.

"What a perfect idea!" Applebee exclaimed. I've never been in an office where there was a rocking chair."

"President Kennedy gave me the idea. His back was injured in the war. Not only that, but my father is a master craftsman when it comes to furniture."

"What a treasure your chair must be. Your nephew and his friend are fortunate to have such a skilled woodworker in the family."

"Giuseppe taught me his crafts as a young boy. You can see that I choose to study law instead of sawdust."

"Applebee spread his arms in supplication. "My father wanted me to become a Psychiatrist. I became an educator instead, who wanted to create minds instead of controlling them."

"We have one thing in common then. We became what we wanted to become."

"Has Romeo and Roberto created a problem?" Patrick asked, changing the subject of rocking chairs and careers.

"Your nephew was involved in a problem with one of his teachers. It also includes Roberto, whose legal guardians are your sister and her husband."

"Why did Winslow's father bring you into this problem with his son?"

Headmaster Applebee wished he had never known Winslow. He knew that Romeo and Roberto spent the afternoon in Winslow's mansion on Webster Street, instead of the library.

"The Merrimac Academy has an advanced Computer that is able to solve a calculus formula, known as the N- dimension. His son, Winslow, is part of a special organization who can reach ...."I cannot explain it!"

"We must look into this." Patrick said. "Once I confront Romeo he will tell me what this is all about. You need to sign these papers on Professor Periodicus. If he ever returns to question us, the Professor will find out that we have a big fat file from his past. He is history."

When the Headmaster left them, Carmen asked him if they ought to have invited Applebee for dinner. "He did you a favor driving the boy's home from school."

"So we could sort out our dirty laundry in front of him?" Patrick replied with a kiss.

CHAPTER FOUR

QUANTUM EXPLAINS THE FOURTH DIMENTION TO ROMEO AND ROBERTO

"Step inside the shower." Winslow instructed them.

"Wait! Hold on." Romeo cried out. "We have to take a shower?"

"No. Trust me you guys. We are going to take a ride in an elevator."

"Oh in that case," Romeo said stepping into the shower, up or down?"

Roberto moved in along with Winslow.

"You don't have to be sarcastic Romeo. I do trust Winslow. You heard the voice in the pyramid."

"You are so naïve Roberto. Showers do not go up or down like an Otis elevator."

The floor of the shower dropped swiftly as the shower descended into the bowels of the earth beneath Winslow's house. Romeo held his breath until they suddenly slowed and stopped. Winslow opened the door and led them out of the shower elevator.

"Awesome!" Roberto cried.

"Magnificent. Romeo added.

"Welcome to my secret room guys. Now it is your room also."

"You told us we were going to climb the stairs." Romeo told him.

"What I meant was we were going to climb the stars.

"Where are we?" Roberto asked.

"You are in the fourth dimension." Winslow told them.

The place they entered could not be called a room. The darkness before them went on forever, surrounding them in a circle like an infinite void, a cosmos of stars, planets and galaxies, extending beyond their imaginations, far into an expanded universe.

They followed Winslow as he guided them through the room that was not a room. Winslow spoke and as he did, they walked into a dome the size of a small planetarium. The top of the dome glistened in a dark blue sky with a warm pleasing light.

"What is this place?" Romeo asked.

"We are in Quantum's control center." Winslow replied.

"We are in a computer control center?" Roberto asked. "Why did you bring us here?"

Winslow gestured with his hand to where three chairs instantly appeared. They seated, themselves facing toward each other.

"Let me explain a few things to both of you that I could not do until you were in this room."

THE DINNER MEETING ON TAYLOR STREET

They finally gathered around the dinner table. "Roberto, you are a great wine steward." Rita told him as he and Romero were seated across from each other. "Romeo, grandfather tells me you answered all of his riddles but one."

"You finally stumped him?" Angelo asked?
"Not even Roberto solved it!" Giuseppe bragged.

"Would you like to try it on the Professor and me?" Andretti asked Giuseppe.

"Riddles are neither proper conversation over dinner nor do I find any humor in them. They are such childish jokes!" Isabelle declared as she joined in."

"Now dear," Giuseppe reminded her. "We are guests here."

The meal was served. Rita's silver tip roast was a perfect medium rare with the ends well done for those who preferred darker servings. Golden brown pan fried boiled potatoes were served with pepper garlic gravy along with broccoli steamed in white vinegar.

"Romeo refused the broccoli until his father reminded him that dessert was chocolate cake. "Can I have his dessert if he doesn't eat his broccoli?" Roberto asked.

"Not unless you want to start a food fight." Romeo told him.

"Then it means you are going to choke down your broccoli. That will be the day you start a food fight." Roberto said.

"How can you let them talk like that Rita?" Isabelle asked.

"They are just teasing. The boys are forever challenging each other Mother."

"It is another improper dinner conversation. Children should never speak at the table unless they need to have someone pass them the salt or pepper."

Following dessert, Patrick stood up and raised his glass. "Well mother, as one of your children I would like to speak. It was an excellent dinner as usual, but tonight Rita and I invited you and father to announce my engagement, to be married to Carmen and for her to marry me. Carmen will now announce another upcoming event."

"What!" Isabelle cried. "Patrick, you can't be serious. "What do you know about Carmen? We haven't met her family yet. How can you do this to your parents without going through the proper traditions?"

"Listen to what Carmen has to tell you mother. Then you will know." Patrick said.

Carmen stood next to Patrick raising her glass as she spoke. "We will be blessed with twins in about six months. We will have two more children in the family."

Romeo and Roberto signed signals across the table. Roberto nodded agreement to what Romeo communicated. "May we be excused mother? Romeo asked. "We have to clean up our rooms."

"Run along boys. You may not want to hear the rest of this."

"We just need to solve another one of the riddles." Romeo told her.

QUANTUM'S CONTROL CENTER

"We would never have believed any of this if you told us yesterday." Romeo told Winslow.

"Right! Right." Roberto agreed.

"Are you afraid?" his friend asked.

"This place is awesome Romeo!"

Winslow laughed and began explaining.

"First, you are in the fourth dimension. Creatures in the third dimension cannot see you. Second, the fourth dimension breaks two of the laws in the third dimension. These are the law of gravity and the law of impenetrable force. Thus you are now defying gravity and the law stating that no two objects can occupy the same place at the same time.

Your bodies are invisible in the third dimension. You are now in a parallel universe where light moves at a greater speed. In the fourth dimension you can travel from Earth to Mars in three earth days measured in third dimension time.

We have the second N-Dimension formula that allows us to enter the fourth dimension. This formula cannot get into the wrong hands." Winslow ended his explanation.

Romeo knew many of the calculus equations and properties of quantum theory from his father's cousin Andretti. Both he and Roberto spent many hours with Professor Xthanlos in the ancient Egyptian libraries, translating the astronomical symbols of ancient Egyptian scholars. Their astronomers were knowledgeable in the field of dimensional calculus. Few quantum physicists exist today working in the field of dimensional calculus.

"Why bring us here Winslow?" Roberto asked. "Who would the wrong hands, you speak of, be?"

"The world governments sponsoring and funding the space programs do not want us to reach the fourth dimension. To do so would thwart their plans to exploit the resources of other planets in our galaxy. They are trying to destroy any attempt to develop Quantum theories and build Quantum Computers."

"The Quantum Computer," Romeo asked. "You have a quantum computer?"

There are only three Quantum Computers on earth. One is located in Van Couver B.C., one at the University of Southern California, and the one in this room." Winslow said.

Slowly, they turned around finding themselves facing a monitor the size of a theater screen. "Romeo and Roberto, meet Quantum. Quantum, these are my new friends Romeo and Roberto."

"Welcome Merrimac students. I've received much information about both of you."

Romeo spoke first. "We are talking to a computer?"

"Romeo we talk to computers everyday on the phone. They tell you what numbers to press or ask you to speak in order to find out what you want to find out. They tell you that you have reached an automated system which is actually a computer." Winslow told them.

"There are only three Quantum computers? What makes them different from IBM's Big Blue Computer?" Roberto asked.

Quantum clicked and hummed. "Permit me to answer for Winslow gentlemen. We three Quantum Computers are equipped to store all of the information in the known universe of the third dimension. In addition, we are able to store fourth dimension calculus. Mathematics is the language of all dimensions. Our power allows us to enter the N-Dimension where we are all sitting now.

The ability to do so requires that our processors be stored underground at the temperature of liquid oxygen, which is 360 degrees below zero."

Romeo and Roberto listened carefully. They said nothing, even though all kinds of questions danced through their minds.

Quantum continued. "I will now explain the dimensions to you."

"What about Professor Applebee picking us up to drive us home from the library?" Romeo asked Winslow.

"You are in the fourth dimension. Headmaster Applebee is in the third dimension. We will be waiting for him in the library. You do not have to measure time in the fourth dimension."

THE HOT TUB – ROMEO AND ROBERTO

"We are safe here." Roberto said. "Whatever are we going to say to Andretti and Professor Xthanlos about Winslow and Quantum?"

"They have always welcomed us in their hot tub. Remember the Greek baths last summer?"

"They were the best bath experience we ever had!" Roberto told him.

"We never knew what our bodies were until we were treated in the baths." Romeo added.

"I would never have known how many touches would please me, especially the bamboo massages."

"Not until we entered the pyramid." Roberto said.

Romeo slid into the water of the hot tub. He sat next to Roberto. "Our minds were touched by a magic voice. Andretti said the voice was one of three voices everyone hears but only listens to two of them."

"What three voices?" Roberto asked.

"Andretti says that they are me, my own self and I."

Roberto placed his arm around Romeo. "Which one of you am I hugging now?" He asked.

"You are hugging me. According to the Professor, I am not me."

"Then who is I?"

"I am everyone who wants me to be what they want me to be."

"What about your own self?" Roberto prodded, as he held Romeo in the swirling waters around them.

"Andretti believes that the self is the one voice that allows us not to become what we don't want to become."

"Like your grandmother?" Roberto said. "She thinks everyone is selfish."

"We better get to bed before they finish their after dinner drinks." Romeo suggested. They showered and walked through the exercise rooms on the third floor. When they climbed into bed, Roberto kissed his friend goodnight. "I know you figured out what the answer to Giuseppe's riddle was."

"My grandfather must be very happy now that he can spend more time in his woodshop making two cribs instead of one. And yes, you know the answer too!"

"Tomorrow, they chimed in unison, it is always coming but when it gets here it is gone!"

"Speaking of tomorrow," Roberto asked, "How can we meet Winslow and his computer, in his secret room? Who will drive us home from school? What if your Uncle Patrick suspects something? "When will we tell Andretti and Professor Xthanlos what is going on? Who is Winslow, really?"

"Remember tomorrow never comes when it is here." Romeo told him. "We left Quantum's control center and found ourselves in the school library. Quantum told us that there were governments sending agents, working for them, to prevent any organization from discovering the fourth dimension. Winslow is protecting the N-Dimension formula from falling into their hands."

"So, they are the hands he is telling us about?" Roberto asked.

"We must believe him."

"Romeo, we are out of our own dimension now. When they put me in that group foster home, I would have ended up on the streets if your parents did not rescue me by becoming my guardians."

"Roberto, when I first saw you in my first grade math class, there was a feeling that we were never ever going to be, never ever, the best of friends."

"I made a great first impression on you Romeo. Why or what made you change your mind?"

"Remember the class bully who threw water in your face? You pulled out your penis and pissed on his shoes. He never came near you again."

"It wasn't water that he threw in my face Romeo. He urinated into a cup in the bathroom before he walked into the hall to toss it into my face. Did you notice he never wore those shoes to school again?"

"Not only that, but when he got near you the next day, you pulled down your zipper. We could hear him crying all the way to the exit doors. It was at that moment that I decided to get to know you better. The class bully was always trying to take my lunch money. Once we were friends my lunch money would be safe. That was how we started to be friends."

THE FOUR DIMENSIONS

Quantum explains the four dimensions.

"In the first dimension we would be standing in one line, like a railroad track. Other people would stand in front of us or behind us in the line. We could only talk to the person in front of us or the person behind us. Our only direct connection would be with them.

In the second dimension we could talk to the person to the left or to the right of us. Our source of information is now doubled as we have two dimensions in which to communicate.

In the third dimension we are like a submarine or airplane. Now we get our information from each other; forward, backward, left and right, up or down. As you know by now, in the fourth dimension you can be in the same place at the same time and not be visible to those in the other three dimensions."

"Cool!" Romeo and Roberto cried at the same time.

CHAPTER FIVE

ANGELO AND RITA

The guests were kissed goodnight. The kitchen was back in order as the dishwasher splashed away. Angelo snuggled next to Rita as they sipped on liquor which Andretti gave them from his summer in Greece.

"What is it about the Greeks"? Angelo whispered. They distill magic brews from special herbs like the Germans do with caraway seeds. You add some ice water and they turn into a cloudy cocktail."

"Maybe we need to add some of this to our sausages." Rita suggested.

"What would the USDA require us to name it as an additive? There would be no alcohol left in the process. We could say we used Rosemary sage and thyme."

"You would use Kummel made with caraway seeds? That means we would add Kummel during the processing of the meat?" Rita asked.

"We must not shop talk tonight," Angelo told her. "We can do this at the Hobbit Farms tomorrow. Our family comes first."

"Why does everything have to be so complicated?" Rita asked.

"Life really isn't, when you get down to it."

"What is it that it is?" Rita asked. "You are talking in circles!"

"My cousin Andretti told me that it is what it is, whatever it is and wants to be, in order to become it." Angelo explained.

"You are even crazier than Andretti." Rita told him.

"You think that Romeo and Roberto are too intimate with each other?" Angelo asked. "You want us to look for ways to keep them apart?"

"The professor and your cousin can take Roberto, this summer, to Greece while Romeo spends his time with us." Rita said.

"What would Romeo do with us? You listened to his poem. Separate them? Our son would never forgive us for doing what you suggest. Roberto is never going to Greece this summer without Romeo. Even though my cousin and the Professor would not interfere with our decision, there is no power on this earth strong enough to separate those boys."

"Angelo, Romero and Roberto have found a new friend at Merrimac. When I ask them about him, they tell me he is helping them with calculus equations. They meet with him in his assigned library alcove. They are hiding something from us Angelo."

"This is what children must do to prevent their parents from forcing them to become what it is, they want them to become."

Angelo was suddenly between the devil and the deep blue sea. "My father wanted me to please my uncle." Andretti told her. He filled his cordial glass and allowed the orange cognac to warm his palette before it descended down, into his toes.

"Your uncle and your father had no idea that you were planning to run away to the circus?"

"We were not only hiding our plan to escape, we just kept it between us as we took off into the night."

Rita placed her wine glass on the table. "We must meet their new classmate."

"Let's invite him and his father here for dinner. Patrick and Carmen can join all of us. That way it won't be uncomfortable for the boys. Winslow's father is one of Patrick's clients."

"We can hire Chef Di Caprio from the Four Spades Tavern to do the catering." Rita added. "Patrick can put it on his expense account."

"In that case we can invite Andretti and the Professor. Patrick will bill us later for at least half of the expenses."

"It will be hidden under billable hours for Hobbit Farms." Rita said. "Patrick and Carmen will have two more mouths to feed once the twins are born."

"We could pay for the dinner from Romeo's college fund. This dinner is so we can meet their new friend." Angelo suggested.

"Angelo. Patrick and I set up Romeo's fund with a proviso that no one, including you and me, could withdraw any funds. Once Romeo has his eighteenth birthday party, only Romeo could decide what to spend it on." Rita explained.

"If we separate them, it is possible that they will run away from us, like Andretti and I ran away from the sausage factory." Angelo said. He imagined that his son would never allow his parents to do what Rita suggested.

"Then I will reserve any judgment until after the dinner with Winslow and his father." Rita promised.

"You send me to the moon!" Angelo said.

"Only to the Moon," She asked, what about the other planets?"

A VISIT TO THE DARK SIDE

Once again Winslow's bathroom dropped into the bowels of the earth. Entering the secret room, Romeo asked Winslow how Quantum was able to draw upon the power and energy needed in order to function and operate.

"The center of the earth has what we call a quantum wave," Winslow explained.

"Like the waves of the ocean?" Roberto asked.

"Yes except quantum waves are composed of the same elements and materials which make up our sun. Earth's vast oceans in their purest form are the mixing of two gases."

"Of course," Romeo said, "Hydrogen and oxygen. The sun does not have enough oxygen to make water."

"Billions of years ago our sun threw off great balls of fire." Winslow continued. "These fire globes kept rotating until they fell into orbit around the sun, to become our solar system."

"Then why didn't our star keep on creating more globes of fire?" Romeo asked.

Quantum Computer appeared on the large monitoring screen.

"Because the globes, which we call the children of the sun, began to orbit their parent source, a planetary system was formed resulting in the creation of an electromagnetic field."

30

"How did this prevent our star from sending out more globes?" Romeo asked.

"Three of the principles of quantum physics are involved." Quantum explained. "One: The sun like every star has a fixed amount of energy. Two: Every atom in our universe rotates. Three: Our star, the sun, was held in place by the new planets, as a gravitational force field developed."

Quantum produced the image of our solar system on the monitor. As the planets rotated, slowly orbiting the sun, Quantum projected the view of electromagnetic force fields on the screen. The planets with the brightest or strongest currents orbited nearer the sun while those with weaker currents orbited at greater distances away from the sun. Notice that each planet has a magnetic field."

"Like a ring!" Roberto cried. "Like the magnetic field on earth."

"The electrical path of each orbit fits inside the ring!" Romeo exclaimed.

"These electrical currents are known as magnetic flows or orbital fluctuation moas. Scientists in the third dimension are just beginning to explore the moas as they begin to understand the creation of matter." Quantum told them.

Roberto was amazed. "This means that our solar system is like an atom in another universe."

Quantum hummed.

CHAPTER SIX

THE WEDDING ON TAYLOR STREET

Patrick and Carmen agreed to spend their honeymoon in Greece. The ceremony would take place in the brownstone garden on Taylor Street followed by the reception. It would be catered by the chef from The Four Spades Tavern.

"Dominic, the owner of The Four Spades, offered to have the bachelor party in the basement of the tavern the night before the wedding. Instead of a naked girl jumping from a large cake, the cast from the Lyric Opera Company agreed to join the party, in the basement, where they gathered each week, to blow off steam. Romeo and Roberto were allowed to attend the party. At first, Rita objected until Angelo explained that other boys would be present during the session to perform on the stage. They would play an instrument or sing in front of the Lyric Opera staff, who could decide their future in the opera world.

"What about the Diva's and the female members of the cast?" Rita asked.

"Well, as you know it is a bachelor party. We would like you and Carmen to entertain the women upstairs in the tavern, until the party is over." Angelo told her.

Rita thought about it. "Why should the females miss out on all of the fun and the music? Is it tradition," She asked, "or the fact that bachelor parties exclude females? What about the stripper who pops out of the cake?"

"She doesn't count?"

"Angelo, we cannot do this to each other. A bachelor party for Patrick cannot be some kind of a guy thing where women are excluded. What kind of a message will we be sending to our boys if we allow this?"

It was Angelo's time to think. "Andretti and Professor Xthanlos would agree with you, one hundred per cent, maybe even two hundred per cent, if that is possible. We must have the party together. How could I be so insensitive? What will your brother think of me for suggesting such nonsense?"

"Patrick is in another world. He is going to be married to a woman who is carrying two of his babies, while he works everyday solving other people's problems."

"Leave the bachelor party to me. It is the wedding ceremony that I'm concerned the most about. My mother will be there, Angelo. If anyone challenges the marriage, or speaks now and forever hold their peace, our mother will be the one to speak up."

"She will find a church doctrine that forbids the marriage of a woman who has conceived."

"What about the virgin Mary. Did she marry Joseph? Angelo asked.

"There is no record in the scriptures."

"So the son of God was a bastard? Angelo asked.

"Did the priests of Jerusalem decide to proclaim that Mary was a virgin? There are many contradictions in the scriptures which finally surfaced with the discovery of the *Dead Sea Scrolls.*
"
"Angelo, we have not decided what to do about Isabelle. Her theology is like a priest's belly button."

"How is Isabelle's theology like a priest's belly button?" He asked.

"When the umbilical cord was cut from the birth mother, the ancient priests proclaimed that the shape of the belly button resulted from the umbilical cord being severed by the one who used a left or a right hand, to separate the child from its mother." Rita said.

Angelo lifted his shirt and pushed his finger into his belly button. "I have wondered how all that lint got in there. Andretti told me that, when we were living in the barn at the zoo, someone was going to help me build a fire when they removed the lint.

"Andretti foresaw this?" Rita asked.

"Yes". Angelo replied. Isabelle and the priests use the same system of logic."

"Then we will allow Romeo and Roberto to join us for the dinner we have prepared for Winslow and his father?"

"We must have their friend spend the night with them. It is called a sleepover. His father will not be here to control him. If we do not approve of this boy then you will speak to Romeo and Roberto."

"Why do I have to be the one who bears bad tidings, should we disapprove of their friend?" Angelo asked.

"A son needs his father to guide him when a boy strays from having friends who lead him astray. It is not something a mother does to her son without him turning against her, Angelo."

"Whose suggestion was it to have Roberto spend the summer with Andretti in Greece, while you kept your son next to the apron strings? Remember, we have allowed our son to embrace Roberto as his best friend. Children are given to us for the time it takes to create, not only their body, but rather to help them develop their minds. We have given a body Rita."

"What about their minds? We teach our children to be honest, to love others, to seek truth and knowledge, above all else Angelo."

"The systems in our bodies are governed by our genetic nature. What you speak of Rita are those matters governed by the heart. "What we become, is a result of our culture education and experience. "The dimensions of the mind have no limit in our universe."

Rita clapped her hands and placed them on each side of her head. "You sound just like your cousin and the Professor!"

"Andretti and I would still be working in Uncle Carlos's sausage factory, if Professor Xthanlos had not rescued us."

"Since Romeo travelled to Greece with them this summer, he has changed. So has Roberto." Rita sighed. "It is difficult to decide what they need to know and who they learn it from. I don't want Romeo growing up too fast and learning new things we don't understand."

Angelo put his arms around her. "What you don't want is why we are together today. We did not enroll our boys in Merrimac Academy to atrophy in some parochial school. They may learn more in one day, with their friend Winslow, than with any teacher at school."

"Patrick says Roberto and Romeo meet beneath a statue of a scarecrow in OZ Park."

Angelo pulled her closer to him. Kissing her passionately, he knew that they both would approve of Winslow. The boys would travel with Andretti and the Professor, along with the honeymooners, to the world of the Mediterranean islands.

DINNER WITH WINSLOW

Romeo invited Winslow and his father to the brownstone for dinner, following the meeting with Quantum. They met in the library.

"My parents want your father to meet my father. There will be all sorts of people there to hide the fact that my parents really want to meet you." Romeo told him.

"Better still, you are invited to spend the night with us." Roberto added.

"A sleepover with you guys? This will be a new experience for me. The only sleepover I've ever had, ended up with two government security agents trying to kidnap me."

"What about the other guys at the sleepover? What happened to them?" Romeo asked.

"Do you really want to hear this story?" Winslow asked.

"Are you nuts? Of course we do. This library is full of adventure stories." Roberto prompted. I'm sure none of them are as good as your story."

"Was Quantum involved in this story?" Romeo asked.

"Let's make a pact." Winslow suggested. "If we have a sleepover at your place, I will tell you the whole story. Your parents may not approve of me."

"How can they not like you Winslow? My father graduated from Merrimac and my mother saved him from an evil uncle. His cousin Andretti and Professor Xthanlos were the reason why I heard voices in the pyramid that led us to you."

"Quantum knows about the pyramids and the voices. Don't forget that those in a higher dimension can communicate with those in a lower dimension." Winslow reminded them.

"Quantum explained that to us Winslow. If we do have this sleepover at my parents place will any secret agents try to kidnap us" Roberto asked.

"Not unless Quantum decides to give you the N-Dimension formula. Until then you are not in any danger of being kidnapped."

"Winslow. We want to understand the formula and the calculus equations that lead us to another dimension."

"It is up to your parents to approve our sleepover. Then, it is up to you guys to decide where you want to go, once you listen to my story."

"Where would we go Winslow once we decide to join you?" Romero asked him.

"You will be able to travel in the fourth dimension to the dark side of the moon."

"Leaving our family here to worry about our disappearance?" Roberto asked.

"Remember what Quantum told you. The speed of light in the fourth dimension allows you to return to the third dimension in the same place at the same. You are able to return to your home on Taylor Street moments after you entered the fourth dimension."

"What happens to the time we spend in the fourth dimension?" Romeo asked.

"The fourth dimension is time. It is unlimited, unlike third dimensional time. When you discover the N-Formula, you are able to travel and enter the fourth dimension. It is being in the same place at the same time in the third dimension."

"This is very heavy Winslow." Romeo told him. "Do you mind if we run this by Professor Xtanlos and Andretti?" Romeo asked. "They got us into this, in the first place, by taking us to the pyramids."

Winslow stood up and shouldered his back pack. "Don't they specialize in ancient Greek history?" Winslow asked.

"Yes. We spent the summer in Greece and Rome with them. They took us to the pyramids in Egypt." Roberto replied.

"My father and I are looking forward to meeting all of your people." Winslow told them. "The sleepover will be a unique experience."

The dinner was a unique experience in itself. Everyone was pleased when the boys steered the conversations toward their school projects. The evening was a great success.

HOBBIT SAUSAGE FARM

The morning following the dinner with Winslow and his father, Rita stormed into Angelo's office. "We have to talk Angelo! "He is too good to be true."

Angelo was sorting through orders and receiving forms. The purchase orders were next to the paid invoices. "You are supposed to audit these." He told her.

"What about our son?" She asked.

"What about our son?" Angelo asked, handing Rita a stack of papers.

"Do we approve of his friend, or not?"

"What is there not to like about him?"

"The sleepover," She replied. "Patrick drove them off to school this morning as if nothing happened. They ate breakfast as if I wasn't even in the room. What do you suppose happened during their sleepover?"

"I suppose that whatever happened in their sleepover is not our concern." He replied.

"Angelo! Romeo and Roberto are in trouble!"

"I will look into this Rita. Andretti will help me."

CHAPTER SEVEN

HEADMASTER APPLEBEE – MERRIMAC ACADEMY

For the second time in one week, Winslow, Roberto and Romeo were invited to meet in the Headmaster's office. The school secretary seated them on the sofa.

"Mr. Applebee will be with you shortly. He wants to talk to all of you." She told them before closing the door to the room.

"That was a great sleepover last night." Winslow said. "Your hot tub was cool!"

"We didn't have to wear togas." Roberto chuckled.

"We sit in the tub as the water jets massage us, talking and listening to each other." Romeo explained. "We talk about what happened to us during the day. Sometimes we don't say anything. We feel closer together when we are sitting, naked without clothes, in a tub of hot water."

"It didn't take you very long to learn how to turn on the valves and adjust the water jets to where you wanted them massaging you." Roberto told Winslow.

"The best part was telling you about the time government agents tried to kidnap me."

"It was during a sleepover too, with two of your friends from physics class." Romeo added. "Quantum saved you by hanging them on the scarecrow in OZ Park."

"What happened to your friends?" Roberto asked.

"We travelled to Oregon, by train, across the Northern states, with Quantum riding in the last train car. Both Quantum and the railcar were in the fourth dimension."

"Then they were invisible?" Romeo asked.

"There is more to your story, isn't there Winslow?" Roberto wanted to know.

Headmaster Applebee entered the office. "What kind of story are you telling now?" He asked Winslow.

"I was telling them about my trip to our farm in Oregon, last Christmas, sir."

"Do you know why the three of you are having this meeting in my office?"

"No sir. You did not send us an agenda." Winslow answered.

"Is it about Professor Periodicus?" Romeo quickly asked, hoping the headmaster did not notice Winslow's sarcasm.

The headmaster sat across from them as he raised his thick eyebrows. "Yes, we have heard from your chemistry professor. He left, when the academy board of directors placed him on a leave without salary. He is requesting a hearing before the board, to be reinstated. Unfortunately, you three must be present at the hearing."

"Will my Uncle Patrick be your lawyer?" Romeo asked the headmaster.

"He cannot be the school's legal counsel because you are related to him."

"We didn't do anything!" Roberto exclaimed.

"That is how the law works. The professor is entitled to a hearing. He claims you boys slandered him by telling lies. Whether it is true or not, the hearing will determine if we were wrong and the professor was right, or if we are right and the professor was wrong."

"What if Merrimac's Director's refuse to take him back?" Winslow asked.

'Your father will not be permitted to sit on the board at the hearing because you are his son. If the board decides against the professor, then he must take his case before a court of law."

"What happens if the board takes him back?" Winslow asked.

"Like my father would say, every time, when I asked him for a BB-Gun," the headmaster told them, laughing, "It ain't gonna happen son."

You mean to say," Winslow exclaimed, "that Professor Periodicus can turn this into a three ring circus, even though we didn't do anything wrong?"

"What does my uncle have to say?" Romeo asked.

The headmaster rose from his chair. "The hearing has been set for next Tuesday. "Your uncle will talk to you and Roberto tonight."

"What about Winslow?" Roberto asked. "He is in on this with us!"

"Your father will discuss Winslow with you at your meeting with your Uncle Patrick, tonight." The headmaster replied.

"That's just great!" Romeo cried as he left the office. "You meet with us here, when we don't know what your agenda is, yet you know the agenda of our meeting with my father and uncle tonight!"

"Who else is going to be at this meeting tonight?" Roberto wanted to know, joining Romeo and leaving the office.

When they left the room, Winslow studied the headmaster as he sank back into his chair. "My father is in Hong Cong sir. What does he have to say about this?"

"He told me that you will work this out with the help of your friends. It is a puzzle to me Winslow how you can be so calm about all of this. At your age, you are supposed to get good grades, go to school dances and give your parents a hard time. Instead, you have an IQ we are unable to measure."

"I don't dance with girls. My mother was a girl when she married my father. She left me with nurses, nannies and boarding schools. My father was a quantum physicist with the National Aeronautics and Space Administration. When I was born, the government funding supported exploration of outer space. Quantum physicists exchanged lab coats for space suits. Both President George Bush and his son were not even aware that we had a space program.

As a student in an advanced Swiss Academy, my father enrolled me in a research program known as the Quantum String Wave Program. I learned dimensional calculus. It was a fourth dimension math. When we returned here, my father enrolled me in The Merrimac Academy of Mathematics and Science. Chicago is where the atomic bomb was developed."

Chicago is the center of commerce, the medical research mecca of the world, preserving our human cultural heritage. Chicago is the water hole of America."

Headmaster Applebee was nonplused.

Winslow returned to his mansion on Webster Street. There was a message from his father. It reminded his son of how important it was, for him to remain neutral in the world of conflict. His father also reminded him that his fifteenth birthday was coming up and asked his son where he would like to celebrate his birthday.

His third dimension was crumbling. It was time to take the bathroom shower to Quantum's lair. Before he stepped into the shower, Winslow removed his toga.

He entered Quantum's Control Center and danced around beneath the monitor.

"Welcome Winslow. Are your hormone's taking over?"

"Quantum, I know you do not spy on me. I spent last night with my new friends, in a hot tub. We had so much fun."

Winslow explained what was going on. "Quantum, please tell me what at am going to do?"

Quantum clicked and hummed.

"You dance very well Winslow. You have kept your body in good shape. Do you think your father would allow you to study ballet?"

"Get serious Quantum. I don't wish to twirl ballerinas around. We are on a mission here."

"Attend the meeting tonight. You and your friends are sure to come up with a plan to stop Professor Periodicus."

"Like enter their meeting tonight in the fourth dimension?" Winslow asked.

"Perhaps, yes."

"Then what do we do?" Winslow asked. "Can you solve our problem Quantum?"

"I will help you solve the problem when you find a solution. You and your friends must come up with a plan to put an end to your problem."

Then you will help us?"

"Let me know what your plan is. Bring Romeo and Roberto here after school tomorrow."

"Will it be OK to attend their meeting tonight if I am invisible?"

"They will be discussing you Winslow. Do you want to know what they will say about you? Your friends may not be able to continue a relationship with you in a new dimension. Your father reminded you to remain neutral during conflict. He also wants to know where you wish to spend your fifteenth birthday."

"Send my father an e-mail from me Quantum. Tell him that I want a dinner upstairs with Romeo and Roberto." Have him invite Rita, Angelo, Andretti and Professor Xthanlos. We will return their hospitality. Oh, and Headmaster Applebee must join us! By then our plan to deal with Professor Periodicus will be underway."

TAYLOR STREET – MEETING IN THE HOT TUB

Romeo and Roberto walked through the shower room, rinsing themselves, before entering the spa. They entered the hot tub, turning on the valves as they directed the jet sprays. Angelo and Andretti walked into the room, stepping into the tub as they seated themselves opposite the boys.

Angelo adjusted his jets toward the spine. Andretti aimed them toward his feet. It was a long lecture day at the university, requiring him to teach from the standing position.

Romeo looked at Roberto as he closed his eyes and gritted his teeth. Suddenly, another set of jet streams turned on, as they watched in amazement.

"Not again! Angelo cried. "Rita will call the maintenance man tomorrow."

"Romeo and Roberto smiled as they gave a high five to each other.

"We are meeting here boys to talk about what is going on in school. The headmaster tells us that you are involved in a dispute with a former teacher who has requested a hearing before the school board."

"We know what happened in class that day." Andretti added. "Patrick has explained the whole story."

"This problem seems to be centered on your new friend Winslow." Angelo told them. Can you tell us more about him?"

"Winslow was only trying to get the chemistry professor off of our backs. We were not late for class. The professor was using it as an excuse to get us into his office. Winslow knew that. In his office, we were told to sit on Professor Periodicus's sofa. He sat next to us."

"Winslow accused him of having sex with his students."

"Then what happened?" Angelo asked.

"We ended up in the headmaster's office." Romeo replied.

"In his petition, the professor is saying that you lied to the headmaster about him having sex with his students."

"It was Winslow who told the Professor this, while we sat on his couch. We never told Applebee anything." Romeo said.

Andretti sank beneath the spa's water.

"Which brings us to a new problem Romeo," His father told him.

"Wait!" Andretti sputtered. "Winslow is not here to defend himself."

"He is not here to defend himself from what?" Angelo asked Andretti."

"Remember Angelo when we ran away together to join the circus?"

"We ended up in Doctor Perkins zoo. You talked me into going north with him to unload a cage full of jungle animals and a gorilla."

"Then we cleaned the displays, until Professor Xthanlos met us beneath the scarecrow in OZ Park."

"Angelo. You were the same age as your son Romeo is now. His friend, Winslow, is the same age as I was when you decided to enter a new dimension. We had no idea where it would lead to." Andretti explained.

"Romeo is not telling us the whole story, are you?" His father asked.

"When Romeo's mother told you about the boys meeting Patrick under the statue of the scarecrow, you knew that they were seeking the same experiences we wanted, that night, when we agreed to meet your teacher." Andretti told Angelo.

Roberto interrupted the Meeting. "Why are we here trying to solve a problem that also includes our friend Winslow?" He wanted to know.

"We are all in this together." Romeo told them as he ducked under the water. He surfaced and sputtered. "This is a problem for us to solve. Andretti is right father. We started it and we will finish it. Winslow introduced us to another friend who will help us, once we have a plan."

Romeo's father decided to allow his son to solve the problem that they were having with Professor Periodicus. "You and Roberto may join with your friend Winslow in finding the solution, Romeo."

"Can we have another sleepover with Winslow?" Romeo asked.

"When we slept in the hay loft, in the barn, at the zoo Angelo, we did not call them sleepovers." Andretti told him.

"Because we were sleeping in," Angelo laughed. "Not over."

"What did you sleep in?" Roberto asked. "We spent the summer in Greece, remember?"

"We have always slept together. You must know by now what happens during a sleep in? "Andretti asked.

Romeo dove into the hot tub.

"We know what happens." Roberto said.

Romeo surfaced. "What about another sleepover with Winslow?" he asked.

His father reached over and pushed Andretti under the water. "Do whatever it takes to get rid of Professor Periodicus."

"We want to go to Winslow's place, after school tomorrow. Roberto and I have a plan to stop Professor Periodicus from taking us to court."

BACK TO THE DRAWING BOARD

They walked into Winslow's bathroom and stripped. "Where are the togas Winslow?" Roberto asked.

"We don't need them anymore." Winslow replied. "We splashed in a hot tub naked. We slept together naked. How is it your mother didn't walk in on us Romeo?"

"I have an agreement with both of my parents. They do not enter my room during sleepovers and I stay out of their bedroom at all times, unless it is an emergency."

"We also agreed not to run through the house naked when they are having company over for a visit." Roberto added.

"What if your father is in Quantum's computer dome Winslow?" Romeo asked.

"My father has only one rule in our dress code. Anything goes when he is not here, otherwise proper attire is required at all meals. He is in Hong Kong today, so not to worry because we are having dinner at your place tonight along with a sleepover." Winslow explained.

"We are going to figure out how to get rid of Professor Periodicus." Roberto exclaimed.

"Slide into the shower." Winslow said. "Now that we have a plan, Quantum is going to help us."

They entered the fourth dimension beneath a parallel universe in another galaxy.

Quantum's monitor came to life as he spoke. "This is where he is staying until the hearing." A room appeared on the screen showing them the hotel room where Professor Periodicus was registered. The professor was taking a shower. The bathroom door was open as steam poured into the room. The professor turned off the water and was seen walking from the bathroom, with a towel draped around his waist. As the mist cleared from the room, he seated himself in front of a lap top computer.

"Your plan is a good one." Quantum told them. "All you need to do now is figure out how you are going to entrap him."

"My Uncle Patrick helped us with the legal things. We must be careful with what we say to the professor. He must believe that he is talking to a young boy, no more than fourteen years old."

"One of you will have to be the decoy in the park." Quantum warned them. Then when he tries to pick you up, the police will arrest him. All of this will be recorded, including his statements on his computer."

"Except for one thing, my uncle said that the boy he meets, in the park, cannot be one of us. Romeo told them.

Quantum clicked and hummed.

"We have a lot to think about at our sleepover tonight." Winslow said.

Winslow viewed the monitor displaying the hotel room. "Quantum is going to hack into the professor's computer. All the evidence will be there to arrest the professor, but there will be no proof!"

"He must be caught in the act!" Roberto told them.

"Caught in the act of doing what?" Romeo asked.

"Quantum is taking care of the evidence. We are the ones who must provide proof." Romeo declared, as he danced around in the dome. "It is time for us to head back to Taylor Street."

CHAPTER EIGHT

COMING OF AGE

The birthday party in the Winslow mansion turned out to be, 'The event never to be forgotten'. The artist creates works, no matter how old they become, from the child they never lost.

Winslow's father pulled many strings for his son.

The banquet hall in the mansion was elegantly set by Chef Di Caprio. He was pleased to provide the service of The Four Spades Tavern cuisine, remembering the day when he catered a birthday reception for one of his son's friends.

His son Jeremy was adopted by an elderly woman who maintained a mansion on Webster Street in the same block where Winslow lived.

The party in the Winslow mansion was a formal sit down dinner. Jeremy, a violinist along with his friend Duncan, an accomplished pianist, were hired to entertain the guests. They were students at the Merrimac Academy.

THE PLAN

Earlier, on the day of the party, Winslow, Romeo and Roberto presented their plan to Quantum in his underground room.

"The professor must meet a boy on line and arrange to meet him." Winslow said.

"Only thing is Quantum, it can't be one of us." Romeo added.

"We have to find another boy." Roberto told Quantum.

"Good work lads. You find him and my processors will handle the rest. The conversations between the professor and the boy will be the evidence you need, once the professor meets him face to face."

"Finding the right person is not going to be easy." Winslow told them.

"One of the students at Merrimac must have been invited to sit on the professor's sofa." Romeo said, hopefully.

"There is one! After the professor left Merrimac, a student in my literature class asked what happened to us when Periodicus took us to his office. He told me there were rumors about what happened to a friend of his on the sofa." Roberto explained.

"Do you think this friend would talk to us?" Winslow asked.

Quantum hummed, clicked, and buzzed.

"Tell me his name Roberto. I have all the blogs of our Merrimac students, stored in my third dimension files, during the past twenty four months. This student and his friend are certain to communicate with each other about what took place on the professor's sofa. It is additional incriminating evidence against Periodicus." Quantum told Roberto.

Roberto hesitated. He looked to Romeo for help as Romeo signaled his approval for his friend to reveal the student's name.

"His name is Jeremy." Roberto told Quantum. "Jeremy Di Caprio."

THE PARTY

The catering associates, as they were called, gathered in the kitchen off of the banquet hall. The hall and the kitchen were off limits to Winslow who promised his father that he would not transpond to either of those rooms, in the fourth dimension.

However, this promise did not include, his friends, Romeo and Roberto. They entered the kitchen as invisible guests.

"Did you ever see so much food, more delicious looking then this?" Romeo asked Roberto.

Roberto was tempted to snatch a butterfly shrimp from one of the trays. He resisted the temptation as Romeo reminded him that they were guests at the party.

The catering crew gathered around Chef Di Caprio. He presented his final instructions to them. "The Beef Wellington will be served following the presentation of the salad. Now, this is most important. The dressings for each salad will be decided by each diner. Our dinner guests know what a good salad dressing is. Should anyone of them return their salad to this kitchen, untouched, you will be sent to scrub pots and pans when we return to the Four Spades Tavern.

"Isn't he a bit harsh?" Roberto asked.

"Jeremy told me about his father. He was imported from one of the Vatican kitchens in Rome where his culinary skills were well known. He was removed from his position in the kitchen, due do some problem with the Vatican priests.

"You know more about Jeremy Romeo, then what you are telling me."

"Uncle Patrick is the attorney for the Four Spades Tavern. He takes care of all their legal work and represents Chef Di Caprio as well. He rescued his son Jeremy who was sent to a group home. It is a story for the hot tub."

"Then Jeremy is going to help us stop professor Periodicus. Does your uncle know about this?" Roberto asked as Jeremy and Duncan entered the kitchen.

The Chef greeted both boys. "You are just in time for dinner." He told them, as he motioned for his crew to be seated around the kitchen table.

"It is time for us to leave and join Winslow's party, in the sitting room, Roberto. Can't I take one of those butterfly shrimps with me? Roberto asked.

"Try it and while you're at it, nab one for me too."

"What about one for Winslow?"

"Sure. Let's take three and get out of here. Winslow is waiting for us in the library." Romeo agreed. "I'll grab some napkins."

MERRIMAC ACADEMY

Uncle Patrick agreed to meet with the boys in the school library. Headmaster Applebee decided to join them and listen to the plan. Winslow's father gave his son permission to print the conversation from the internet, recorded by Quantum. Jeremy and his friend, who were invited to sit on Periodicus's sofa, joined them in the plan to take the professor down.

"This still seems like entrapment. Here we are, conspiring against a man who claims that he is innocent. With all of this evidence, on the computer, he can meet this student in person, smell a trap, and decide to abort his own plan." Patrick advised them.

"The Professor has gotten away with his pedophilia for many years, moving from place to place, school to school, even changing his name." The Headmaster told them. "He preys upon young boys. It is our duty to stop him."

"But how can we do this Uncle Patrick?" Romeo asked him.

"We need to catch him in the act." Patrick replied.

"You mean Jeremy or his friend, have to go to the hotel room and let the professor touch them, like he did on his sofa?" Roberto asked.

"The police would arrest him before he could do anything." Patrick told him.

"Suppose the professor asks you to take off your clothes?" Roberto asked.

"We will ask him to take off all of his clothes." Jeremy said. We went through this before, with him, in his office. He stripped and danced around the desk."

"What happened next?"

"We dressed and left the office."

"Nothing else happened?" Patrick asked.

"Yes. We both got an A in chemistry.

Headmaster Applebee was furious. "He gave you an A without learning anything about chemistry?"

Jeremy laughed. "We know more about chemistry sir, than the professor knew. The A was earned on his couch."

"Are you boys willing to do this, so that the professor will not be able to hurt other boys with his problem?" the Headmaster asked.

"No sir," Jeremy's friend replied, "We just wanted to put our clothes back on. The professor never hurt us. He confused us."

"How is that?" The Headmaster asked.

"Jeremy and I were recycled as foster kids. When Jeremy's Grandmother took us to live with her, we knew what it was like to be hurt and abused. When Professor Periodicus asked us to take off our clothes we realized that he cared for us. Jeremy and I figured out what he was doing. He wanted someone to love him exactly as we wanted someone to love us."

QUANTUMS PRINTOUT

SUPERMAN-X:　WHERE DO YOU TWO GUYS LIVE?

BEENYBOYS2:　WE LIVE IN CHICAGO.

SUPERMAN-X:　WHAT AREA?

BEENYBOYS2:　LINCOLN PARK NEAR THE ZOO.

SUPERMAN-X:　COOL! MY PLACE IS NEAR THERE.

BEENYBOYS2:　MAYBE WE CAN VISIT YOU.

SUPERMAN-X:　HOW OLD ARE YOU BOYS?

BEENYBOYS2:　JASON IS 13. I AM MARTIN AGE 14.

SUPERMAN-X:　DO YOU LIKE OLDER MEN?

BEENYBOYS2:　ONLY IF THEY HAVE A GREAT BOD AND A LARGE YOU KNOW WHAT.

SUPERMAN-X:　I WOULDN'T DISAPPOINT YOU! WHERE DO YOU GO TO SCHOOL?

BEENYBOYS2:　MERRIMAC ACADEMY OF MATHEMATICS AND SCIENCE.

SUPERMAN-X:　YOU MUST BE GOOD STUDENTS. A FRIEND OF MINE TEACHES THERE.

BEENYBOYS2:　WHAT DOES HE TEACH?

SUPERMAN-X:　IF I TELL YOU CAN YOU KEEP IT A SECRET?

BEENYBOYS2:　YES. WE KEEP ALL KINDS OF SECRETS. WE WILL KEEP YOUR SECRET.

SUPERMAN-X:　HE TEACHES CHEMISTRY AND HE LIKES BOYS.

BEENYBOYS2:　WHEN CAN WE VISIT YOU? IT WOULD HAVE TO BE AFTER SCHOOL.

SUPERMAN-X:　BOYS, YOUR AGE, ARE VERY HOT. HOW ABOUT TOMORROW AT 4:30?

BEENYBOYS2:　WHERE DO WE MEET?

SUPERMAN-X:　UNDER THE STATUE OF THE SCARECROW IN OZ PARK.

BEENYBOYS2:　GOOD PLACE. RIGHT ON! TELL US THAT YOU ARE NOT A COP.

SUPERMAN-X:　I AM NOT A COP. TRUST ME. ARE YOU COPS?

BEENYBOYS2:　WE ARE NOT COPS. JASON AND I ARE LOOKING FOR SOME FUN.

SUPERMAN-X: WHAT KIND OF FUN?

BEENYBOYS2: WHAT A DUMB QUESTION MAN. WEREN'T YOU OUR AGE ONCE?

SUPERMAN-X: STAND UNDER THE STATUE. IF I DON'T LIKE WHAT I SEE FORGET IT!

BEENYBOYS2: NO SUPERMAN-X. YOU WILL NEVER FORGET US. SEE YOU TOMORROW.

OZ PARK BENEATH THE SCARECROW

Jeremy and his friend Martin walked from the academy steps into Oz Park.

"Don't they play soccer here?" Martin asked. I have never kicked a soccer ball. My father said it was a sissy sport. That was the word he used."

"What sport did you play before coming here?" Jeremy asked.

"My father wanted me to join the football team. He told me that it would look good on my college resume."

"Why didn't you play football then?" Jeremy asked.

"I saw no reason why I should play a sport where the object was to hurt another player. The school football team, were jocks who were taught to inflict pain on the players opposing them. There were always injuries where bones were broken and heads injured."

"What about the equipment the players used to protect themselves?" Jeremy asked Martin.

"They used the equipment designed to protect them, by inflicting greater damage to their opponents."

"Like the ancient warriors of Rome who's armor protected them as they fought before the emperors. Only the best gallant slave warriors, who were forced to fight in the arena, were allowed to live at the end of the battle, as the emperors lifted their thumb. The slave would live to fight another day." Martin told him.

"That is, until he was defeated? How fair was that? Jeremy asked.

"If the warrior slave survived three of his challenges, he was set free."

"You are saying Martin that you fought three sword fights against your father and he set you free?"

"The swords were not weapons that he used against me. My mother died. He raised his own sword against me. My computer saved me."

Jeremy waited while Martin settled down. They were approaching the Scarecrow.

'You never told me any of this, Martin. We are going to meet Professor Periodicus. Group homes suck! They are a real bummer.

"At least Professor Periodicus cared for us." Martin said. "Isn't it wrong to do this to him, to turn him in?

"It is not a question of right or wrong Martin. Like Headmaster Applebee said, the professor must not be allowed to continue using his students and other boys to satisfy his sexual desires. Your father had no right to pierce you with his sword."

"He told me that it was the love of Jesus coming into me."

"It is called rape Martin. It is what happened to me in the group homes. When I told my social worker, he took me to another foster home. The older boys took turns with me until I ran away. They finally found me a home with an older lady who cared for me."

"Then what happened?"

"The Children's Protection Agency decided she was too old to care for a foster child. By then, she had hired teachers who gave me piano and violin lessons. On the day when the case workers arrived to take me away, I was with Emilio, my violin teacher, who made arrangements with his friend Count Sebastiane to hide me at his place in the Hancock Building on Michigan Avenue."

"I really want to hear the rest of this story Jeremy, but look behind you."

Professor Periodicus walked toward the boys. He sat on a bench, across from them. Looking around the park, he feigned a surprise look. "Whatever brings you here?"

"We are waiting for Superman-X." Jeremy said.

"Who are you waiting for?" Martin asked.

"Beenyboys2."

"You found us." Jeremy told him.

"Now what do we do?" Martin asked.

"I like what I see!" The professor exclaimed. "How do I know this is not a set up?"

"Set up?" Martin asked. "What is a set up?"
"The cops are using you to arrest me."

"We are not cops professor. Martin and I have been on line, in the gay room, looking for someone who would have fun with us. We thought you were Superman-X." Jeremy told him as he rubbed his crotch. We saw Superman's poster on the wall in your office. You enlarged the bulge between his legs. We want to see you dance around naked again."

"You both dressed in my office bathroom and left, while I was dancing."

"You were so excited you did not hear the bell ring. We did not want to be late for our literature class." Jeremy explained.

The professor stood up and instructed them to follow him. "What is this stupid statue of a scarecrow all about?" He asked them as they walked toward his hotel.

QUANTUM COMPUTER CENTER

Winslow, Romeo, Roberto and Quantum were silent as they sat in the dome. Above them, the fourth dimension circled in a parallel universe.

"Are you ready to see what is going to happen in Professor Periodicus's hotel room?" Quantum asked.

"No Quantum," we will wait until Jeremy and Martin tell us their story in the hot tub." Romeo told him. "We cannot see anything that happened in the hotel."

"Really," Quantum asked, "Then I will tape what goes on when the professor takes Jeremy and Martin into his room."

CHAPTER NINE

RUB A DUB RUB – FIVE BOYS IN A TUB

Patrick arranged for the depositions to be taken in his law office. Professor Periodicus was arraigned and remanded to jail, without bail. His attorney requested that his client be freed on bail. The presiding judge, upon learning the client's history of forging a new identity, decided to set the professor's bail at one million dollars. His attorney argued for a lessor amount, maintaining that his client was not a violent man.

"My client cannot possibly come up with that kind of money your Honor."

"The bail is ordered at one million dollars   Are there any other motions?" The Judge asked.

"Yes your honor. We request that the petition filed by Professor Periodicus to the Board of Directors of Merrimac Academy, be denied. We have filed the documents with this court. In addition, your honor, we request that the students named in this petition be excluded from any future prosecutions against the professor." Patrick pleaded.

"What does his attorney have to say about this?" His Honor asked.

"My client claims he was wrongfully accused by these students. He claims to have met two of his former students in OZ Park who wanted to go with him to his hotel room. They remembered how kind he was to them, during their chemistry course."

The Prosecuting Attorney objected. "Your Honor, we found the defendant naked in his hotel room, with the two boys, dancing in front of them."

"Did the boys have their clothes on?" The Judge asked.

His Honor called for a court recess.

The party's in the hearing reconvened, taking their places, as the judge entered.

"This is a most unusual, and I will admit, a difficult case before me. First, the petition before the Merrimac Board of Directors is denied. The students named in the petition are not to be included in any future litigation with the defendant.

As to the charges against this client, his acts are a misdemeanor, not a felony. His bail is therefore set at five hundred dollars, plus court costs.

Patrick jumped from his chair at the table. "Your Honor, this man should not be on our streets!"

"Under the circumstances, I have to agree with you." The Judge told him. "Do you have any other motions counselor?"

"Yes your Honor. The Board of Directors at Merrimac Academy would like a psychiatric evaluation of Professor Periodicus. Although he is not a felon, according to our legal standards, he does have issues with young boys. Is it possible to send him for an evaluation?"

The Judge looked around the court room. He set his eyes on the Professor's attorney. "Another hearing would have to take place before an evaluation is ordered."

"My client agrees to an evaluation." His attorney told the judge. The professor did not wish to return to his jail cell.

"So be it." The judge ordered. "Prisoner will be taken to the psychiatric ward at county hospital for evaluation, where he will be held until his trial date."

THE DEPOSITIONS

The attorney representing the professor was present, along with the Assistant Attorney in the Prosecutor's office. Martin and Jeremy were interviewed separately. Patrick cautioned both boys to answer each question truthfully. The depositions were completed in two hours.

The Professor's lawyer remained in Patrick's office once the other attorney's left. He plopped himself on the sofa. "They don't have enough evidence to convict my client on a felony charge. Your boys are not going to give the prosecutors enough evidence for a conviction. You know that don't you?"

Patrick decided, at that moment, that he would consult with Andretti. It was time to gather everyone into the hot tub for a dolphin party. Romeo and Roberto were not telling the whole story.

Rita and Angelo agreed to the party and a sleepover for five boys.

They were happy to accommodate Romeo's new Merrimac friends. Romeo was coming out of his shell. Roberto would no longer be his only friend. Her decision to separate them for the summer was no longer necessary. Angelo silently approved

Patrick and Andretti sat on opposite sides of the hot tub. The boys adjusted the water jets and settled down between them.

The meeting was somber with only the churning of the water breaking the silence.

"Winslow, Patrick began, "Is there anything you would like to tell us?"

"Like what sir?"

"Whatever it is you want to say Winslow, especially your thoughts about the court hearing."

"Thank you for getting the Academy off the hook Sir. My father and the Headmaster will be grateful."

"What is going to happen next?" Jeremy asked.

"Will he go to prison?" Martin asked.

"Only if the Prosecutor can prove that a felony was committed. If so, the professor can plead guilty. In that case the judge can sentence him to a term in prison. If he pleads not guilty the case must go to trial in front of a jury." Patrick explained.

"What would make it a felony?" Jeremy asked.

"The statutes or laws are not clear when it comes to sexual abuse. In this case, the professor would need to perform an illegal sexual act on another person who is a minor.

The court would require the testimony of the minor as a witness against him."

"What if the boy liked the sex act?" Martin asked.

"The law still makes it a felony. It is illegal to break the law."

"It looks like the professor is going to get away with it." Roberto said.

"One of you must testify against the professor during his trial." Patrick pleaded to them.

Andretti intervened. "It is time to initiate the new dolphins!"

Patrick understood that it was time to end the meeting and begin the party. Both he and Andretti exited the tub room. The dolphins took over.

MERRIMAC ACADEMY OF MATHEMATICS AND SCIENCE

Winslow waited in his library study room. Each student had his own corral, as they were called, to study, read and do research work from the volumes located in the stacks. This morning Winslow was sound asleep with his head resting on the table. Romeo and Roberto sidled quietly into the corral trying not to awaken their friend.

They sat on each side of him as Winslow's forehead used his arms for a pillow. Romeo silently counted to three with his fingers. On three, they blew softly into his ears. Slowly, raising his head, Winslow opened his eyes. Without turning his head he spoke.

"Is this what you guys call a blow job?" He asked.

"Best one you'll ever get." Roberto told him.

"I seriously doubt that. However, it was better than what my alarm clock does."
With that, he put his head back down on the table. "At least my alarm has a snooze button."

"Winslow. Wake up. We have some important matters to resolve." Romeo told him. He sat up, leaning back on his chair.

"You are breaking a rule! Roberto cried. "Is it a misdemeanor or a felony?"

Rubbing his eyes, Winslow replied, "It is called an infraction. Once we turn eighteen we will be called felons. Right now we are juvenile delinquents."

"Can't we be called misdemeanerors?" Romeo said.

"There is no such word Romeo. After last night I'm sure that Webster's New Dictionary will include that one in their new words of the year."

"We had no idea about Jeremy and Martin. Uncle Patrick invited them. It was his idea. It must be the lawyer inside of him." Romeo explained.

"What if the name for us, is Dolphinators? Roberto asked.

"You know what that rhymes with, don't you, Roberto?"

"OH! Right you are."

Winslow shuffled them from his corral. "It is time to meet with Quantum again. We must figure out how to send Professor Periodicus to LA- LA land. Try to stay awake during classes. This time we are taking Martin and Jeremy, with us, to meet Quantum. Meet at the Scarecrow after school?" Winslow asked.

"What if they freak out?" Roberto asked.

"Anyone who can survive a sleepover with us, Roberto, is immune to earthquakes, tornados and freak outs." Romeo declared.

"Yes Romeo. We are the Dolphinators! Both Jeremy and Martin survived adversity. They are still dreaming. No one tells them who they are as they reach for the stars."

"Today they will see another dimension." Winslow continued, as he left the library corral. Jeremy and Martin will help us protect Quantum's N-Formula."

"Protect the formula from what?" Romeo asked.

"Not from what, Romeo, rather protect it from whom."

"How do we get Jeremy and Martin into your shower elevator Winslow?" Romeo asked.

"Quantum has sent each of them a message. They will be there."

Winslow left them standing outside of their literature class, hoping to stay awake during his calculus session.

"What do you suppose Quantum told them?" Roberto asked, sitting alongside Romeo.

"How about, 'Give up hope all ye who are about to enter my shower."

SCARECROW – OZ PARK

"This is becoming one of our favorite spots." Jeremy observed.

Martin sat under the statue. "Not just our favorite spot. Winslow sent a message to meet Roberto and Romeo here."

"He is in a good place to meet others. It is in an open space where people can see us." Jeremy told him.

"Like the professor, who liked what he saw, before he walked over to meet with us." Martin added.

"Where do you think we are going?" Jeremy asked.

"To Winslow's place across Webster Street," Roberto said. Then we take a ride in an elevator to his secret room under the mansion."

"How did you two walk up behind us like that?" Jeremy asked.

"Easy, you were looking up at the Scarecrow." Romeo replied. "Winslow is expecting us unless he fell asleep again."

Winslow welcomed them into the mansion. Jeremy noticed that the front hall was very large with a high ceiling. Statues stood along both walls leading toward a wide stairway. He took them to the bottom of the stairs. "Wait here please. Be right back." He then disappeared into one of the rooms.

The four of them sat at the bottom of the stairs.

"What about all of those statues we just passed?" Martin asked. "They were all naked."

"They were nude Martin. The Greeks and the Romans thought the nude body was a beautiful work of art." Romeo told him.

"Like the naked dolphins in the hot tub last night." Jeremy explained. "Underneath our clothes we are all naked."

"Winslow said the same thing the first time we were here." Roberto added.

They turned around at the sound of Winslow's voice. He spoke to them from the top of the stairs. "If you climb the stairs we can introduce Quantum to our new guests."

"Is this going to be weird or scary?" Martin asked Romeo.

They joined Winslow on the second floor at the end of another long hallway. Instead of statutes, the walls displayed beautiful photographs between each doorway in the corridor, pictures of planets, constellations, and moons.

Winslow stopped in front of a photograph of the earth's moon. They entered a bathroom as Winslow closed the door behind them.

"Is this the secret room?" Martin asked, huddling behind Jeremy.

Roberto put his arm around Martin. "No Martin. This is the way to the secret room. We must step into the shower. It is an elevator that takes us down to the secret room."

"Common' Martin, if you can't trust these guys after last night, who can you trust?" Jeremy pleaded.

"We have been to the secret room Martin. I promise you that once you go there you will never be afraid of anyone again." Roberto explained.

"Not even my father?"

"Especially, not even your father." Roberto promised.

No one objected to removing their clothes before they entered the shower. "Won't Quantum, this computer we are going to meet, object to five naked boys standing inside of the secret room?" Jeremy asked.

They stood together as the elevator descended into the earth. Once again everyone was overwhelmed and awe struck when Quantum greeted them. The theater screen size monitor, turned on, showing dolphins rollicking in an island cove as they frolicked and played together. Quantum had broken the ice.

"Like us in the hot tub last night Quantum! "Winslow laughed.

"How did you know what we were doing last night?" Romeo was not afraid to ask.

Quantum clicked and buzzed.

"Winslow will begin." Quantum said. "We have two new travelers who have entered our fourth dimension."

Winslow invited Jeremy and Martin to be seated before the monitor.

"Quantum will explain why you have been chosen to enter the secret room. The dimensions will be explained, along with the reasons why we want both of you to join us. If you decide that you do not wish or desire to enter the fourth dimension, we will understand."

"Why did you choose us?" Jeremy asked.

"You were selected for three reasons." Winslow explained. "One; you read books. Two; You know about mountains. Three; You have had a traumatic experience."

PROFESSOR PERIODICUS

Quantum explained to them how he logged and stored third dimension information. "Winslow did not want to view what happened in the hotel where the professor took you. Here is a recording of what took place with both of you in his room." He was speaking to Jeremy and Martin. "It is like you told the prosecutor. You can see the professor dancing around, in front of you, naked."

"He was telling us the truth, in the chat room, about the size of his penis." Jeremy said, laughing.

"A misdemeanor is the only charge the prosecutor has against him. One of you must testify as a witness, someone who the professor has had sex with. Martin, that someone is you." Quantum told him.

Martin looked into the top of the dome. He raised his arms into a universe he had never seen before. Wrapping his arms around his shoulders, clasping his hands on them, Martin cried.

"He did not pierce me with his sword. He put it in my mouth where it did not hurt. My father never made me do that. He said it was a sin for a man to put a penis into another man's mouth. The professor gave me an A for giving him a blow job."

Jeremy opened Martin's arms, hugging him, until Martin stopped sobbing.

Everyone, including Quantum, waited. They took turns hugging Martin. Quantum opened a new screen. Five dolphins turned into five boys swimming in a hot tub.

Martin raised his fist. "OK! Periodicus is history Quantum."

THE TRIAL

Professor Periodicus was released from the hospital. The judge ordered another hearing in his court. Martin was on the witness list for the prosecution. Patrick arranged for another deposition with the professor's attorney. He insisted that it be taken in his office on Taylor Street. The defense attorney objected, insisting that it be conducted in his office.

"Since the boy is represented by the academy's attorney, there is no reason to conduct the disposition in a new setting." The Judge decreed. "The witness is not an adult. He is a boy, whom I will not allow the defense to intimidate."

"The trial date has been set for next Tuesday." The Judge told them. "Are there any objections?"

The professor's attorney stood. "Your honor, this does not give the defense sufficient time to look into the background of the witness."

"The witness was named at the arraignment and in the first hearing." The Judge explained. "The district attorney has given you due notice counselor. The trial will proceed as directed. Are there any additional motions before this court?"

"Yes your honor. The defense has not given us their witness list." The prosecutor told him.

"What say you defense attorney?" The Judge inquired.

"We will have our witness list in the district attorney's office by noon tomorrow, your honor."

The professor paid his bail and signed the agreement to cover court expenses if the jury found him guilty. He was released, as the Judges gavel ended the hearing.

The trial would begin next week.

THE SECRET ROOM – WINSLOW AND QUANTUM

"Martin is going to testify against the professor Quantum. He agreed with you, to do it, when we met here last time." Winslow said. "There is something wrong about Martin doing this, Quantum."

Quantum buzzed and hummed.

"You did not click! I know that you are trying to figure something out. Martin is in serious trouble, isn't he?"

"Trust me Winslow. "We are all in trouble."

Quantum's monitor opened. "This is the evaluation interview, in the hospital, with Professor Periodicus. He is speaking to his psychiatrist."

Doctor Animus: Do you know why you are here?
Patient: The judge sent me for an evaluation.
Doctor: Is there something wrong with you?
Patient: No. Some of my students got the wrong idea when I tried to help them.
Doctor: You danced naked in front of them professor. The police caught you doing this in your hotel room. Do you believe that this is normal behavior?
Patient: Yes. I have danced naked in front of boys for many years.
Doctor: Is that all you do in front of them. Do you have sex with them?
Patient: Only if they want me to.
Doctor: Have you performed any sexual acts with your students?
Patient: No. Never! No boy ever agreed to have sex with me.
Doctor: What about Jeremy and Martin?
Patient: They required help solving problems with chemical equations.
Doctor: You tutored them in your office after school?
Patient: They both earned an A grade in chemistry.
Doctor: How commendable professor. You must be proud.
Patient: Jeremy and Martin were good students.
Doctor: What about complaints with boys in your past Professor?
Patient: Whenever I have worked closely with one of my students, there have always been rumors and innuendos. None of my students ever testified against me.
Doctor: According to the Judge who sent you here Professor, there is one of your students listed, as a witness against you.
Patient: Which student?
Doctor: I am not at liberty to say professor. You will have to talk to your attorney. You are being released from the hospital tomorrow morning for another hearing before the judge.
Patient: What about your evaluation report?
Doctor: Good luck professor. You may return to your room.

THE PROFESORS HOSPITAL ROOM

Bruce the Moose welcomed the professor back to their room. "How did it go with the Doctor, Mr. Innocent?"

"They have a witness who will testify against me!"

"You better have a good attorney. If the jury believes the boy, professor, you are toast." Moose told him.

"What would you do Moose?"

"Find out who the boy is and deal with it. He is a kid who is living with his mother. The father is not around. Then, find a way to keep him from testifying."

Professor Periodicus began packing his bag. "Thank you Moose, I will miss you.

"Don't let that boy send your ass to prison professor. Remember, a dead person cannot testify against you."

"What are you saying Moose?"

"If you end up inside the prison walls, they will take turns with you. Many convicts were sexually abused as children."

"I never abused my students! "

"Try to explain it to your fellow inmates, professor. If the jury doesn't believe you, can you expect your cellmates to pat you on the shoulder? They will label you as a sexual predator, a pedophile preying upon his students."

"Moose, what am I going to do?"

"Did you, or did you not, professor have sex with your student?

The professor zipped his travel bag. He did not answer the Moose.

"There you are." The Moose told him. Their witness did not appear against me, at my trial. When the doctor evaluates me, I am a free man."

"At my hearing tomorrow, I will plead not guilty and be free on bail." The professor said.

"Free to change what will happen to you professor, if you do the right thing. You cannot change yourself, so do the right thing." His roommate advised.

IN THE SECRET ROOM

"Moose is going to be set free?" Winslow asked.

"Like Napoleon he is going to meet his Water Loo." Quantum replied.

"Moose is going to be exiled on an island near Sicily?"

Quantum buzzed and hummed. "Good to know that you paid attention in history, Winslow. No, Moose will be sentenced to life in a hospital for the criminally insane."

"Did he kill the witness in his last trial?"

"He tried to, but the Moose was followed every place he went. The boy he was stalking turned out to be a police officer disguised as a young boy. When Moose approached the officer with a sharp hunting knife, he was shot in the leg, the kneecap to be exact. Moose fell to the ground." Quantum reported.

"Do you think Professor Periodicus took Moose's advice?" Winslow asked.

"What do you think Winslow?"

"You told me that we were all in trouble Quantum."

"As the Ancient Scriptures advise us, "Our human mind is fickle. There is no telling what it will do if given the right time and the place. "You have viewed the evaluation interview between the professor and his doctor. In addition, we recorded the conversation with the professor and his roommate Moose."

"Do we all have delusions Quantum?"

"Yes Winslow. "Delusions are able to make us stronger, helping us to solve our problems. Trying to fool others requires us to deny that our problems do not exist. We cannot solve a problem until we know what the problem is."

"Quantum, what is our problem now, other than to protect Martin from the professor?"

"Professor Periodicus was sent to Merrimac to steal the N-Dimension formula."

"How can he do that Quantum?"

"He can kidnap Martin and Jeremy."

Winslow opened his cell phone. "I will Message everyone for a meeting in the library, tomorrow morning Quantum. We must not allow the professor to meet with Martin or Jeremy before tomorrow's hearing."

Quantum hummed and buzzed.

"Is the professor from the National Defense Agency?" Winslow asked.

"No. They are a rogue outfit working with the CIA. There is no record of them or how they are funded."

"Are there any more of their agents looking for us?"

"There is one more. He happens to be the professor's attorney, who knows that Martin is the witness, testifying against the professor. They call him Agent Hercules. The rogue outfit is known as Zeus." Quantum explained.

"My friends need your help Quantum. What do we do?"

"It is time for all of them to learn the basics of traveling in the fourth dimension Winslow. Romeo and Roberto did well at your birthday party. No one suspected anything when they snatched the shrimp and napkins. Bring everyone here after school tomorrow. Once the dolphins are settled following their elevator ride, they will be able to experience travel in the N-Dimension. In addition to my instructions, you will see who followed you from the Academy to the statue of the Scarecrow."

"What about from the Scarecrow to Webster Street?"

"Need you ask, Winslow?"
"All of us will be invisible after we leave the Scarecrow? What if Jeremy and Martin notice that we are invisible?"
Quantum buzzed and hummed.

SETTING THE TRIAL DATE

When the Judge was seated, everyone in the courtroom took their places.

"Will the defendant please rise." The Judge requested.

Professor Periodicus and his attorney stood.

"We waive the reading of the details leading to the charges. Attorney Hercules told the judge.

"To the charge of misdemeanor one, how do you plead professor?"

"If it pleases the court your honor, my client has a name."

"So let it be known in the records, counselor. How does Professor Periodicus plead to the charge?"

"Not Guilty your honor."

The judge continued. "To the charge of sexual abuse with a minor, a class one felony, how do you plead Professor Periodicus?"

"Not guilty your honor."

"Prisoner is released from custody. The bail is set at 500 dollars, plus court costs. Trial is set for next Tuesday at ten o'clock, AM. Professor, I have reviewed the results of your psychiatric evaluation. The attorney for Merrimac Academy has been granted a restraining order against you and your attorney. The students involved in this case cannot be approached or contacted by anyone connected with you or your attorney. To do so would result in a serious contempt for this court."

WINSLOW HAS A PLAN

"The Judge has no inkling of who these rouge agents are Quantum."

"They are beyond any power or control, authorized by the judge, to stop them."

"We can stop them Quantum." Winslow promised. My plan will succeed."

Quantum buzzed and hummed. This time he clicked.

"You also have a plan to help us, don't you Quantum?"

"See you dolphins after school tomorrow. There is much for me to do tonight Winslow."

CHAPTER TEN

ROMEO DECIDES TO CHALLENGE ANGELO AND ANDRETTI

Giuseppe knew that meeting his grandson, without Isabelle's knowledge, could end up in another one of his trips to the basement, playing another game of solitaire. The two cherry wood cribs for Carmen and Patrick's twins, were almost finished. Romeo's mother decided that Romeo needed to spend more time with his grandfather. Her son never stayed overnight with Isabelle and Giuseppe. It was time for Romeo to sleep over with them.

Isabelle refused to accept Romeo as her grandson. He was conceived in sin. Her religion did not sanction abortion. Isabelle pondered the problem with several vodka drinks.
Romeo's mother must abort her bastard child. Rita gave birth to a boy, who agreed, years later, to spend the night with his grandfather, over the objections of his grandmother.

Giuseppe roused Romeo from his sleep. They crept down the back stairs to the kitchen. His grandfather poured Romeo a glass of orange juice.

"Drink this with your juice. Take one every day."

"What is this grandfather?"

"All of the vitamins and minerals you need to keep you alive Romeo."

They went to his basement work shop. He watched his grandfather turn a block of wood on a machine. He called it a lathe. Using a chisel, he turned it into a beautiful table leg."

"You can do anything you want to, if you have the right tools and a good eye." He told Romeo.

Giuseppe hugged him. "Remember Romeo, the right tools, a good eye and someone who loves you."

MERRIMAC ACADEMY LIBRARY

The school librarian called Headmaster Applebee who was seated on his office sofa, enjoying a coffee and bagel covered with cream cheese.

"There a ruckus in one of the corrals sir."

"You mean an uproar or fight?"

"More like a lot of noise sir, something we don't permit in our library. You must come at once!"

"Don't you think you can handle this alone?"

"You mean call campus security Mr. Applebee? "I cannot deal with such noisy confusion."

"Right, you have never been married with children running around the house."

"Heaven forbid! Taking care of books is more than enough for me sir."

"Understood Mr. Librarian, I'm on my way. Maybe you should call the National Guard to protect us before I get there."

"You're joking, right Mr. Applebee?"

DETENTION

The students sat on Mr. Applebee's sofa. "Is everyone comfortable?" He asked them.

"Yes sir." The boys replied together.

"Are you going to eat that cream cheese bagel?" Roberto asked.

Romeo laughed and elbowed his friend. "How about sharing it with me?"

The Headmaster swiped the bagel from the coffee table and put it on his desk. "Because of you delinquents, my coffee is cold and this bagel is mine." He told them.

"We were just joking." Roberto explained.

Winslow, trying to hide a smile, spoke to the Headmaster. "It was my fault Mr. Applebee. I sent them a message to meet in my corral this morning. We were going to get together after school. The librarian freaked out when he saw all of us in my corral. We didn't make any noise until one of us knocked over a chair."

"It was me!" Martin cried. "Send me to detention."

"I'm afraid that the fallen chair is not the issue here. The rules clearly state that no more than three students can occupy a corral at the same time. The librarian was upset."

"We were only trying to get together after school." Martin explained.

"Then you were successful in doing so." The Headmaster agreed. You will report to the librarian after school. He will supervise your detention until five o'clock."

THE LIBRARY – THE LIBRARIAN – THE DENTENTIONEES

They arrived, for the second time that day, in the library. The next 90 minutes of their lives, were in the hands of the Librarian. Seating themselves at the reading tables, the students opened their back packs, removing their homework assignments.

The Librarian suddenly loomed over them. "Oh, no, no, no," He said. "In my detention class you do not do homework assignments."

"Detention is a class?" Winslow asked.

"Students have a lot to learn when they misbehave. The discipline in this library will make all of you wish for mercy."

"What kind of mercy?" Romeo asked.

The librarian held a book in his hand and raised it above them. "This is the Holy Word of God. Never doubt any word written within the holy scriptures of this book." He demanded, placing a copy of 'The Holy Word of God' in front of each student.

"Has Merrimac become a seminary?" Winslow asked him.

"Never you mind," The librarian said, God did not create mathematics and science!"

"Then why do they exist?" Romeo asked.

"The answers are all here in the scriptures. The students in Merrimac Academy cannot possibly imagine what God's plan is, for all of us."

Martin raised both hands. "You are telling us, Mr. Librarian, that God's plan is written in this book you have given us?"

"The Holy Bible before you is the only true word of God."

"Then why," Martin asked, "are there hundreds, even thousands, of books in our library explaining mathematics and science?"

"Your detentions will continue for the rest of the school term, until all of you agree with me."

The librarian reminded them that there would be no more freedom from detention until he released them. For now, you will open your books to Genesis: Chapter one."

"Winslow!" Romeo whispered. "Is it true what he says about the detention program?"

"In a way, yes, the librarian can hold us here until Applebee agrees with him."

"Quantum is waiting for us Winslow. What are we going to do about getting out of here?"

NARROW ESCAPE

They met in front of the Academy.

"What happened?" Jeremy asked.

"Quantum did this." Martin replied. "Are we going to be in more trouble with Applebee?"

"Our librarian is going to wake up tomorrow wondering where we are. Quantum sent a copy of our detention tape to Applebee. His preaching days are over." Winslow said, laughing.

"What is so funny about that?" Martin asked.

"The Librarian will work with the boys on the soccer team. His new position will allow him to discover the more important kicks in life." Winslow explained.

"Like what?" Jeremy asked.

"By not making us believe all the stories in his Holy Book, as if they really happened."

"Like Jonah being swallowed by a whale or a snake giving an apple to Eve." Martin told them."

"What about Lots wife being turned into a pillar of salt? Roberto added. "Then water was turned into wine by a man who was born to a virgin."

"The same man who walked on water and turned a dead man back to life."

"Like our librarian, the priests have always enjoyed their metaphors." Romeo said.

"I don't understand." Martin told them.

"It is not important to understand Martin. Think about it. It is more important to love." Winslow tried to explain.

Romeo interrupted the conversation. "We need to cross OZ Park. The Scarecrow is waiting."

"Isn't the Scarecrow where all of this started?" Roberto asked.

"In more ways than one," Romeo replied." Lead us on Winslow!"

QUANTUMS ROOM

Once they entered Quantum's computer center, they were assigned individual stations with their own keyboards and monitors. Each monitor displayed a video of them, as they were crossing OZ Park, moments earlier, escaping from the librarian.

"Who are those two guys following us?" Jeremy asked"

"They are wearing black navy coats and watch caps pulled down over their ears." Roberto said.

"Wait until you Dolpinators reach the Scarecrow." Quantum told them.

The boys walked beneath the Scarecrow and vanished from the screen. Their stalkers removed the woolen watch caps, staring at the Scarecrow in amazement.

"It is Professor Periodicus and his lawyer!" Martin exclaimed. "Why are they following us?"

"That is why we are here Martin and not in the library. Quantum will explain everything, I hope." Winslow assured him.

Quantum did explain everything to Martin and Jeremy. Quantum told both boys how they could move from one dimension to another. The N-Dimension formula was in danger of falling into the wrong hands. Both Jeremy and Martin were chosen to help keep the formula safe from the governments rogue organization. The professor was using them to steal the N-Dimension equation.

"When they could not hack into Merrimac's Computer or kidnap Winslow last year (Read Winslow's Secret Room), they decided to use both of you to help them."

"How were they going to do that?" Martin asked.

"They needed to find out a way to get all of you, together. They planned the trial in order to reach Winslow and the formula." Quantum explained. "Martin is a material witness. They are going to prevent Martin from testifying. Their plan backfired on them."

"No!" Roberto cried. "The only way to keep Martin from telling on the professor is to keep him from the trial. What can we do Quantum?"

"If any of you are threatened Roberto, I will transpond you to this room, where you will be safe. Just say 'Morning Glory'. Can you remember that?" Quantum asked them.

CHAPTER ELEVEN

THE BROWNSTONE - FAMILY MEETING ON TAYLOR STREET

Gathering in the sitting room for their weekly meeting, Rita and Angelo served the refreshments when Carmen and Patrick arrived, joined by Professor Xthanlos and Andretti.

"The wedding arrangements are complete." Carmen announced.   "The ceremony in the garden will be followed by a reception, catered by none other than Chef Di Caprio."

"What about the honeymoon?" Andretti asked.

"The tickets and reservations, for three weeks of romance in Greece and Rome, are taken care of." Patrick replied. "Our ship sails from New York as soon as the semester at the university ends."

"You are travelling with Andretti and the Professor along with Romeo and Roberto?" Rita asked.
"Once the ship arrives in Athens, Carmen and I will head for Rome." Patrick explained. "Then all of us will meet again in Athens where Carmen and I will fly back to the states."

"What about the trial?" Angelo asked.

Before he could reply, Romeo and Roberto burst into the room.

"Can't you boys enter a room quietly?" Rita asked.

"Why?" Romeo asked.

"We were discussing something important."

"You didn't tell us that Grandpa Giuseppe and Isabelle were coming to our family meeting? They just pulled up outside." Romeo told her.

Rita jumped from her chair. "What on earth are they doing here?"

"Maybe they came to visit us?" Roberto guessed, trying to hide his toothy smile.

Patrick went into the hallway toward the front door of the brownstone, to greet his parents. Giuseppe and Isabelle entered the sitting room. Everyone stood.

"Take your grandfather and grandmother's coats Romeo." Rita kissed her father on the cheek, leading them to the sofa. "Roberto will serve you your drinks."

Giuseppe placed his coat over Romeo's shoulders, handing him Isabelle's fur wrap. Uncle Patrick escorted his nephew to the hall closet.

"We have to talk." Romeo whispered.

"Let's go up to my office. They won't miss us with all the confusion your grandmother makes."

Patrick sat in his office rocking chair while Romeo settled next to him on the sofa. "It's about the trial with Martin and Jeremy."

Romeo told his uncle what happened to all of them during the school day.

"How did you escape from detention?"

"When the librarian handed us the bibles, he excused himself to go to the bathroom. That's when we ran from the library. While Winslow was taking us to his place we stopped beneath the Scarecrow. There were two men following us, Professor Periodicus and his attorney."

Roberto entered the office. "They are calling for both of you downstairs."

"Can you get in touch with Martin and Jeremy right away?"

"Sure." Roberto told him. "They are both waiting in our bedroom."

"What are we going to do?" Romeo asked.

Patrick thought for a moment. "Act like it was any normal day at school. Meet me in the hot tub after dinner."

"Can Winslow be there too?" Roberto asked.

"Is he also in your bedroom Romeo?"

"No. But he can be in the hot tub when we are."

ROBERTO THE WINE STEWARD

Giuseppe raised his glass of Chianti. "Isabelle, why don't you tell everyone, the reason for this surprise visit? Roberto, please compliment my glass with another of your fine wine."

"What is this reason, mother. Are you entering a convent?"

"Don't be sarcastic dear. Giuseppe and I wish to adopt Roberto."

The wine steward entered the room. Giuseppe's glass of Chianti crashed to the floor, upon hearing Isabelle's wish.

RITA AND ANGELO

"Thank you dear for cleaning up the mess and getting my father another glass of wine. Where did Roberto run off to after he heard my mother?"
"He and our son have locked themselves in their rooms. When Giuseppe and Isabelle left, Patrick told me what happened to the boys in school today. The two boys involved in this trial are staying with us tonight. Patrick believes they are in danger."

Angelo told her about all of them getting detention in the library, the escape from the librarian and being followed by their former chemistry professor and his attorney.

"Romeo and Roberto are not supposed to be involved in this trial. How did they get mixed up in this?" She asked.

"Patrick says the boys know more than they are willing to tell him."

"You and Andretti must tell Romeo to stay out of this!" Rita insisted.

"We are having a hot tub meeting tonight. I'm sure Patrick will talk some sense into them."

"Romeo will listen to you and Andretti. He has always followed your advice before"

"Any advice we have given to Romeo was always based on what we would do Rita. Telling our son that he cannot help his friends would not come from our hearts. Wait until tomorrow morning when it all cools down."

"What do you have to say about my mother and father adopting Roberto?"

"What a surprise! You are the one who wanted to separate Roberto from our son. This might be the perfect opportunity to have both your mother's and your wish come true."

"I only want what is best for Romeo. You know that Angelo."

"We can only make certain Romeo has a good education with positive experiences. It is Romeo who must decide what is best for Romeo."

"It isn't easy. Our son will soon be a teenager. What will happen when Patrick and Carmen have twins?"

"Professor Xthanlos told me that one of the mistakes parents often make teaching their children, is not realizing how much their children teach them."

"Sounds just like him. "You know that I would never allow my mother to adopt our Roberto."

"We have always been of the same mind since our eyes met that day in the office at Hobbit Farms. It was the first magic moment, of my life, when you turned me into a love slave."

"Well my love slave, after your hot tub meeting, your master will be waiting for you."

"Will there be candle light?"

FOUR DOLPHINS IN A TUB

The meeting in the hot tub tonight followed a day of mystery upon mystery. Romeo led everyone from his room to the hot tub room while the adults were having dinner downstairs.

"What did you tell your mother?" Jeremy asked Martin.

Roberto turned on his water jets. Soon the tub was alive with currents and bubbles.

"She said I could spend the night here again." Martin told them.

"Everyone thinks that my friend Duncan and I are staying with my grandmother." Jeremy said

"Where is Winslow?" Roberto asked.

"When we left his place, this afternoon, he told me there was some unfinished business to do," Romeo explained, "that we were to start without him."

"What do you suppose it was?" Martin asked.

"Winslow has some kind of plan to solve Martin's problem." Romeo told them.

"My father, Uncle Patrick and Andretti will be here soon. This is not going to be a friendly get together. My grandmother Isabelle announced tonight that she and Giuseppe want to adopt Roberto. We will never be allowed in the library again. Two men are following us. Martin is in danger, and we haven't had dinner. What else can happen?"

Angelo and Andretti entered the hot tub room followed by Uncle Patrick. They took their places in the water opposite the dolphins.

"Angelo spoke first. "Considering everything that is happening Romeo, your mother and I want you to stay out of this problem with Jeremy and Martin."

"Is that what you want also?" Romeo asked Andretti.

"They only want is best for you." Andretti advised.

"What about you Uncle Patrick?"

"Do you have any idea what Martin is going through, the courage it took for him to become a witness! How can you do this to me! Even worse, how can you do this to Martin?"

"Please son, this is for your own good." Angelo said.

"You are not telling us everything that is going on Romeo, especially with your friend Winslow." Uncle Patrick explained.

"Do you want me to avoid seeing Winslow also?"

"We think this would be a good idea since all these problems began when you met him." His father said.

"What are you going to do with Martin and Jeremy?" Roberto asked.

"We will keep an eye on them." Uncle Patrick said.

"To make sure we stay away from them?" Romeo asked.

"To protect them until the trial is over." His uncle replied.

"Roberto is going to stay with Giuseppe and Isabelle for a six week trial period Romeo."

Another set of jets opened in the hot tub.

Angelo cursed. "I forgot to call the plumber! That has got to be fixed."

The four dolphins could not hide their smiles.

"You are grounded for two weeks Romeo," his father told him. "There will be no more visits with Winslow. Roberto will be going to Isabelle's right after school tomorrow. You can help him pack his bags tonight."

"You have never grounded me father. What is it you and mother are afraid of? Do you think I will run away from home like you and Andretti did?"

"You wouldn't get very far in today's world Romeo."

"Maybe not very far, but close enough father."

"Morning Glory," Romeo said as he disappeared from the tub.

"Morning Glory," Roberto said as he vanished.

Jeremy and Martin shrugged their shoulders. "Morning Glory", they both said leaving

the adults alone in the hot tub.

(To be continued in Volume 3)

Books by J H Trembley

Winslow's Secret Room – ISBN 9781477464144 - $5.95
The Book With a Hole In It Vol. - 1 – ISBN 9781477484678 - $9.95
The Book With A Hole In It Vol. - 2 – ISBN 9781478367833 - $9.95
SUNFLOWER CHRONICLES Book 1 – ISBN 9780615650852 - $12.95
The Secret Room
SUNFLOWER CHRONICLES Book 2 -- ISBN 9781477594827 - $12.95
The Farm
SUNFLOWER CHRONICLES Book 3 - ISBN 9781477608784 - $12.95
The Dark Side Of The Moon
Romeo and the Orphan of Hamlet – Vol. 1

ISBN

9781480163041 - $9.95
Now available on Amazon.com/books
And Amazon.com/Kindle
J H Trembley
jhtrem@hotmail.com